Space Telescopes
Patrick H. Stakem
Number 22 of the Space Series

(c) 2018

Number 22 in the Space Series

Table of Contents

Introduction...5
Author..7
Exploration...9
How do we get the data back?..11
How do we control those orbiting Observatories?.................13
Sensors and Detectors..15
    Radiant energy sensing and measurement.......................15
        Star Sensors 17
The Stars, our Galaxy and others, the Universe, and all that. 18
    Astronomy...18
    Our solar system – what is there to observe?..................18
        Sun...19
        Mercury...20
        Venus...20
        Earth..20
        Near Earth Objects..21
        Mars...22
        The Asteroid Belt...22
        Jupiter..23
        Saturn...23
        Uranus...24
        Neptune...24
        Pluto and beyond..25
        Kuiper Belt..25
        Dwarf Planets...26
        Comets...27
        The stars 27
    From somewhere else...31
    Exoplanets...31
Space Weather...34
Astronomy Missions...35
Airborne Observatory...36
NASA's Astronomy satellites...36
    HETE...38

- Swift 38
- Fermi 39
- Uhuru 40
- Small Astronomy Satellite, SAS-2 40
- OAO 41
- HEAO 41
- IRAS 42
- NuSTAR 43
- RXTE 43
- WISE 44
- COBE 45
- AMS 46
- WMAP 47
- Sampex 48
- IBEX 48
- SDO 49
- Ulysses 50
- IXPE 51
- UVC 51
- IUE 51
- IBEX 53
- EUVE 53
- IRIS 53
- GALEX 54
- CHIPS 54
- FUSE 55
- Nancy Grace Roman Space Telescope 55
- Parker Solar Probe 56
- RHESSI 58
- SWAS 58
- WIRE 58
- WIND 59

Alexis (LANL) 59
- Radio Astronomy Explorer 60
- POLAR 60
- MSX 61

NASA's Great Observatories 61

- CGRO 62
- Chandra 63
- Spitzer 64
- HST 65
- JWST 68

Other Nation's Space Telescope Projects 71
Some Planned Missions 72
Observatories on the Moon 74
- International Lunar Observatory 74
- Virtual Observatories 75

Search for Exoplanets 75
- Kepler 76
- TESS 77

Waypoint-1 78
Wrap-up 78
Glossary of Terms 80
Bibliography 94
Resources 98
If you enjoyed this book, you might find something else from the author interesting as well. 101

Introduction

This book covers the topic of Telescopes in Space. Observation of the night sky's started way before Civilization. Early humans looked at all the lights in the sky, saw that some moved, and some were fixed. What did it all mean?

The initial assumption was that the Earth was the center of the Universe, and everything traveled around it. But, by observations, some "heavenly bodies" wander about in the sky, and increasingly complex theories were developed to address this. Later, when the Earth was dethroned as the center of the universe, it was understood that other planets and the Earth were in orbit around the Sun. And, the Sun was just another star, like those that could be seen in the night sky.

Early civilizations such as the Babylonians, Chinese, Greeks, Egyptians, the Maya, and many more kept detailed records of the movement of the celestial bodies, in order to develop a theory about what it all meant. With enough data, eclipses and lunar phases could be predicted. The North Star, or the Southern Cross could be used for navigation. Calendars could be produced, and the phases of the moon understood. You knew when to plant crops, and when to harvest them. In Egypt, you knew when the Nile inundation would come. No more guess work. The learned classes or the priesthood understood these phenomena, and were making sense of the world. Well, they still thought the Earth was flat, and the center of the Universe, but their models worked well enough for civilization to advance.

When you look through a telescope at a distant object, you are looking back in time. You are seeing the light that left the object a long time ago. Atmospheric distortion is a problem, partially addressed by adaptive optics, better addresses by putting the telescope above the atmosphere.

The latest approach is to put the telescopes in space, beyond the interference of the atmosphere. Now we can see more clearly, and farther. The atmosphere not only causes scintillation, but filters higher frequencies such as the ultraviolet, x-rays, and gamma rays. It we want to see in these wavelengths, we need to get up above the atmosphere. Essentially, the atmosphere has "windows" in the optical and radio frequencies. Using high altitude aircraft and balloon-borne telescopes is a partial solution.

This book discusses some 45 space-based astronomy missions. It is not comprehensive. The discussions are sparse, only covering the basic details. The good news is, all of these missions are in archives on the Internet, starting with the NASA sites. This information is included in the resources section, at the end. The really good news is, most of the data from these missions are also archived in open repositories.

There is no particular order to the missions discussed. I have tried to group missions by objective, but that is not always possible, as some missions observe in multiple regions of the spectrum.

Looking at this as a whole, it is interesting how much time and effort have been put into these projects. Once we get a good view, beyond the atmosphere, we start tackling the big picture issues. The biggest one is, are we alone in this Universe?

The observations depends on the state of the atmosphere. Radio telescopes, working at lower frequencies than the IR, visible light, or UV telescopes, work fine on Earth, but have to be located in a relatively radio-quiet zone. Examples are the National Radio Astronomy Telescope in Greenbank, WV and the radio telescope at Aricebo, Porto Rico. This was constructed in the 1960's, in a large natural sinkhole. It is one thousand feet in diameter, and was the largest such facility in the world for 50 years.

It suffered a catastrophic collapse, but is being rebuilt. The Chinese have since built a larger one. The Arecibo facility is listed on the U. S. National Register of Historic Places. It does have a visitor's center. It can also serve as a transmitter,

This book covers the descriptions and missions of NASA telescope missions, and describes some upcoming ones. It includes missions in which other entities collaborated with NASA. This is a new and updated edition, following the successful launch of JWST.

A large set of definitions and acronyms are presented. You can't fly an astrophysics mission without a good acronym.

Author

The Author received his first telescope while in grade school, and was hooked on exploring space from then on.

Mr. Patrick H. Stakem has been fascinated by the space program since the Vanguard launches in 1957. He received a Bachelors degree in Electrical Engineering from Carnegie-Mellon University, and Masters Degrees in Physics and Computer Science from the Johns Hopkins University. At Carnegie, he worked with a group of undergraduate students to re-assemble, modify, and operate a surplus missile guidance computer, which was later donated to the Smithsonian.

He began his career in Aerospace with Fairchild Industries on the ATS-6 (Applications Technology Satellite-6) program, a communication satellite that developed much of the technology for the TDRSS (Tracking and Data Relay Satellite System). He followed the ATS-6 Program through its operational phase, and worked on other projects at NASA's Goddard Space Flight Center including the Hubble Space Telescope, the International Ultraviolet Explorer (IUE), the Solar Maximum Mission (SMM), FUSE, some of the Landsat missions, and Shuttle. He was posted to NASA's Jet Propulsion Laboratory for Mars-Jupiter-Saturn (MJS-77), which later became the *Voyager* mission, and is still operating and returning data from outside the solar system at this writing. He initiated and lead the international Flight Linux Project for NASA's Earth Sciences Technology Office. He is the recipient of the Shuttle Program Manager's Commendation Award, and has completed 42 NASA Certification courses. He has two NASA Group Achievement Awards, and the Apollo-Soyuz Test Program Award.

Mr. Stakem has been affiliated with the Whiting School of Engineering of the Johns Hopkins University since 2007. He supported the Summer Engineering Bootcamp Projects at Goddard Space Flight Center for 2 years, and several cubesat summer courses.

Exploration

Exploration from Earth orbit takes two forms, looking down from orbit are the weather guys, and the climate guys. Looking up are the Astrophysicists. This book is about looking up, trying to make sense of our solar system, galaxy, and universe. We want to be warned of and avoid extinction events. We want to know if we are unique, or have neighbors.

Astronomy dates back far in the history of mankind, starting with observations of the sky, and the development of theories of what it all meant, and how it all worked. The initial assumption was that the Earth was the center of the Universe, and everything traveled around it. But, by observations, some "heavenly bodies" wander about in the sky, and increasingly complex theories were developed to address this. Later, when the Earth was dethroned as the center of the universe, it was understood that other planets and the Earth were in orbit around the Sun. And, the Sun was just another star, like those that could be seen in the night sky.

A lot of exploration takes place from Earth-based observatories, in the optical and radio frequencies. These observatories have to look through the atmosphere, which is blurry. But, the method has been used for thousands of years, and many major discoveries have been made. Another giant leap was made when photography replaced direct observation through the telescopes. Wouldn't it be better if we could put the telescope in space? Why, yes, it would. Many observatory satellites have been placed in orbit, the best known being the Hubble Space Telescope.

As an interim measure, an aircraft or balloon can take a telescope above most of the Earth's atmosphere, and remain there for hours. The first balloon borne astronomy mission was Stratoscope I in 1957. The Boomerang (Balloon Observations Of Millimetric Extragalactic Radiation And Geophysics) project looked at the cosmic background microwave radiation. It flew above 125,000 feet. First flight was in 1997. In 1998 and 2003, it ascended from McMurdo Station in Antarctica. The mission duration on those flights was 2 weeks. Another approach that has been used is to have a balloon carry a sounding rocket to altitude. These can be launched from anywhere, on land or water, and are still in use

The High Energy Focusing Telescope was a balloon payload looking for hard x-ray sources, in 2005. The High-resolution gamma-ray and hard X-ray spectrometer (HIREGS). flew from McMurdo Station, in Antarctica in 1991 and 1992.

NASA's Galileo Airborne Observatory was hosted on a Convair 990 aircraft. It was first used in 1965, but was unfortunately destroyed n a mid-air collision in 1973.

NASA's Kuiper Airborne Observatory used the larger

Lockheed C-141 Starlifter. It operated from 1974 to 1995.

The Joint NASA-DLR Stratospheric Observatory for Infrared Astronomy (SOFIA) has been in use since 2010. It uses a Boeing 747SP with a 2.7 meter telescope. The aircraft is based at the Armstrong Flight Research Center in Pasadena, CA. The telescope is specially mounted to isolate it from movements of the plane. Missions are flown at 42,000 feet, which allows observation of Infrared above most of the atmosphere's blocking water vapor.

How do we get the data back?

The Ground Segment of an orbiting mission consists of the antennas, RF receivers and transmitters, and communication lines to the Control Center. The Ground segment is the link between the control center, and the satellites. All data are archived at the ground segment, in case of communication interruption with the control center. Before high speed data lines, this was done locally on magnetic tape, which would later be shipped to an archive.

For NASA missions, the infrastructure is in place. Originally, the tracking stations were linked with the GSFC facility over leased dedicated landlines. Data communications was provided by NASCOM, using 4,800 bit blocks. Today, almost all of the traffic moves over the Internet.

The Near Earth Network (NEN) communicates with near-Earth orbiting satellites (out to Lunar orbit). It uses the NASA ground stations. There are two stations in Florida, in proximity to the Kennedy Center launch site, at the launch facility at Wallops Island, Virginia, and at the McMurdo Base in Antarctica. In addition, other commercial ground stations can be used, under contract to NASA. Goddard Space Flight Center in Greenbelt, Maryland, manages the NEN, which was formally know as the Ground Network (GN). The dish antennae of the GN range from 34 meters to 70 meters in diameter.

The Space Network (SN) dates back to the early 1980's, when NASA introduced a constellation of communications satellites to replace the ground tracking stations.

The Tracking and Data Relay Satellites, in geosynchronous orbit, are the Space Segment (SS) of the Space Network. They implement communications between to low Earth orbiting spacecraft, and one of the TDRS ground segments. The ground segment units are located at White Sands, New Mexico, and on Guam Island, in the Pacific. White Sands also serves as the controlling station for the TDRS spacecraft. The TDRS network was declared operational in 1989. STDN stations at Wallops Island, Bermuda, Merritt Island (FL), Ponce de Leon (Florida), and Dakar, Senegal, remained operational.

The Tracking & Data Relay Satellite System is over 30 years old, and is being refreshed with new technology. The Space Segment has spare assets in orbit in case of failure.

How do we control those orbiting Observatories?

We need a Control Center! A satellite control center has a wide variety of task. It provides services to the mission 24x7x365. These services include the reception, archiving, limit-checking, and conversion of the received telemetry data. Today, received telemetry is archived in raw form, and saved in engineering units in a database. Use of a standard commercial database simplifies operations and controls costs. The Control Center disseminates selected data to users, either located in a control room, or via the web. The Control Center is usually built around a STOL, or scripting language, for automation of operations where possible. The software does limit checking of incoming data, and issues alerts if limits are exceeded. The control center also is responsible for commanding the spacecraft.

The Goddard Space Flight Center is the hub of the NASA world wide communications Network, and the Lead Center for unmanned spaceflight. It was dedicated in 1959 by rocket pioneer Dr. Robert Goddard's widow. The Goddard center has worked on hundreds of spacecraft projects, including the Hubble Space Telescope, and the new James Webb Space Telescope.

The Control Center provides work space, data systems for telemetry, command capability, and coordination of activities between subsystems. The Control Center is responsible for spacecraft planning and scheduling. It is operating 24 x 7. The Control Center becomes a busy place during spacecraft anomaly or failure situations. It is the workplace for the flight operations team. Generally, the Control Center has a front room, and a back room. The front room is responsible for all real time activities, and commanding. The back room is for off-line analysis. Different personnel may inhabit the back room at different times and mission mission phases. In the early days of spacecraft, the control center function was done from the launch site, or a tracking station. As spacecraft computers become more powerful, some of the functions of the control center could be done onboard. In addition, the control center architecture has evolved from mainframe computers, to the client-server architecture, to pc's, and then to distributed models, based on the cloud. It is important, though, to still have a defined control center "space" where the operations team can congregate. Modern control centers allow engineers to access data on their smartphones.

Satellite Control centers for NASA non-manned astronomy missions are located at NASA's GSFC in Greenbelt, Maryland. The science data may also be processed there, or sent to another site.

The GSFC Visitors Center is located off of IceSat Road, which is a turn off of Greenbelt Road. The Main Gate (employees only) is at 8800 Greenbelt Road, and Icesat Road is East of the Main gate. After turning on Icesat Road, make the next left turn into the Visitor's Center parking lot. Admittance is free. There is a gift shop and bathrooms. Down the hill is Bldg. 14, NASA's communications nerve center, NASCOM. A must see is the awesome *Science on a Sphere* presentation, which takes about 45 minutes. They have a Hubble and a James Webb telescope model.

Sensors and Detectors

The spacecraft carries a series of sensors and detectors, optimized to the particular spectrum or energy level we want to examine.

Detectors tell us of the presence or absence of a signal. If the signal is present, we can derive data about it with sensors. Sometimes we are looking for transient events like a gamma ray burst, or we might be monitoring ongoing phenomena like the Solar Corona.

A *sensor* is a device that measures a physical quantity by changing state in response to the stimulation, and producing a signal. It is an analog world. It is rare that we get to interface directly to a digital source. Some sensors may indicate one of two states (presence/absence) with a simple digital signal that may only require voltage level shifting.

Radiant energy sensing and measurement

This includes the spectrum of radio frequency, infrared, optical, ultraviolet, gamma rays, and beyond. The particular sensor technology used depends on the frequency, but the principles are the same.

Thermal detector devices transform a radiant, infrared or

thermal transfer energy stimulus into an electrical signal. Most often, the absorbed energy stimuli causes a change in the detectors temperature, and this change in temperature manifests itself as a change in electrical resistance or electrostatic polarization. The sensitivity is limited by the fluctuation in energy of the absorbed energy.

The measurement of radiant energy has many applications. The choice of the measuring device depends on the frequency range of the energy and the particular application location or structure. Thermal detectors include thermocouples, bolometers, thermal imaging devices, resistance-temperature devices such as doped germanium and silicon cells, optical pyrometers, photoconductive cells and photoelectric cells, voltage-current devices such as a thermistor or Seebeck Effect devices. Some of these radiant energy detectors require special operating temperatures.

Photodetectors are semiconductor devices that can detect optical signals through electronic processes. Quantum photodetectors are nonequlibrium devices. A minimum amount of photon energy is required to create a quantum excitation; this is referred to as a photo-absorption threshold. The photodetector output signal depends on the non-equlibrium between excitation and recombination in the semiconductor crystal lattice. The signal amplitude is directly dependent on the photo-excitation lifetime and the number of absorbed photons.

Star Sensors

Star sensors measure star coordinates in a spacecraft frame of reference and provide attitude information when these observed coordinates are compared with known star directions obtained from a star catalog. Star sensors can achieve accuracy's in the arc-second range. Most star sensors consist of a Sunshade, an optical system, an image definition device which defines the region of the field of view that is visible to the detector, the detector and an electronic assembly. The detector such as a photomultiplier transforms the optical signal into an electrical signal. Solid-state detectors may be noisier than photomultipliers. The electronics assembly amplifies and filters the electrical signal from the detector. If the amplified optical signal from the detector is above a fixed signal intensity, an output is generated signifying the star's presence.

A charge transfer device star sensor is an optical system consisting of a digitally scanned array of photosensitive elements whose output is fed to an embedded microprocessor. A charge pattern corresponding to the received image of the star field viewed is produced. The charge pattern is then read out serially line-by-line to an analog to digital converter and this subsequent signal is stored in memory.

An interferometer superimposes waves, and uses the

resulting interference to gain information. In theory, this works at any frequency. In practice, sometimes it is difficult or expensive. The technique is useful for the measurement of small displacements. We can combine the outputs of two telescopes, giving a resolution equal to the separation of the two sensors This is a common technique in astronomy. In some cases, different telescopes on different spacecraft, very far apart are used. The Space Interferometry Mission was planned to use this technique for locating habitable exo-planets, but was the victim of Budget cutting.

The Stars, our Galaxy and others, the Universe, and all that.

Astronomy

Astronomy is the science that focuses on the celestial sky, and the objects visible in it. It is based on physics, mathematics, and chemistry. It is the basis for our understanding of the Universe, and our place in it. Besides, its a chance to get out at night and see the pretty lights.

Keep in mind, we can observe not only in the visible light spectrum, but on either side of that – infrared, ultraviolet. We can observe in the radio frequency realm, for Gamma Rays, and even Gravitational waves

Our solar system – what is there to observe?

Our solar system consists of the Sun, which is our source

of power, and a collection of planets, their moons, a large asteroid belt, a collection of comets, and a bunch of miscellaneous rocks. It is a very diverse system, with examples of just about every object imaginable. Mankind has been observing and documenting the solar system for thousands of years. The behavior of the "fixed stars" and the "wanderers" (planets) were carefully documented. Many theories of the solar system and the universe in general were postulated. It was obvious that the Earth was the center of the universe, and modeling that universe was extremely difficult. The ancient Greeks were able to untangle the data enough to predict eclipses. It was not until the late middle ages when Copernicus suggested that the Sun might be the center of our solar system, which made the model much simpler, but almost got him thrown out of the Church.

Sun

The Sun is the source of our energy, and it keeps the solar system organized. It is about a million times the size of Earth. The Sun is extremely hot due to nuclear reaction. The tiny amount of its energy output that strikes the Earth makes life here possible. All the stars you see at night are other suns that are very far away. Each of these might also have planets. Planets around other stars have been discovered from observation, and by their gravitation effect on their sun.

Each planet in our solar system is in an orbit around the Sun. The time it takes a planet to complete an orbit is what we call a year. Each of the planets have a different duration year length. A planet also spins like a ball as it continues in its orbit. The time to complete one revolution is called a "day."

Mercury

Mercury, the closest planet to the Sun, is in tidal lock with the sun, with one side always facing it. It wobbles a bit, creating a twilight zone that is much less extreme. It has no known moons, or Trojans. Being so close to the Sun, it is difficult to observe the planet and its immediate vicinity.

Venus

Heavy greenhouse clouds trap the solar energy, and cause massive global warming on a planetary scale. The surface temperature is high enough to melt some metals. We need to find out what went wrong on Venus, and try to avoid that on Earth.

Venus' atmosphere is 96% carbon dioxide at a surface pressure of nearly 100 times Earth's, a greenhouse gone wild. It has no moons. Venus is roughly Earth-sized, but something went terribly wrong

The heavily clouded atmosphere makes it difficult to observe Venus.

Earth

We observe the Earth continuously with a series of orbiting satellites, to keep track of the weather, and violent events. In fact, Earth is the best observed planet, and we know the most about it. Not that it can't give us surprises. Earth's moon is the largest object in the sky, and has been observed since humans looked up. It is orbited by a series of satellites, has surface landers, and has been visited by Astronauts from the United States.

The bow shock is plasma from the Sun hitting the Earth's or other planet's magnetosphere. The plasma is ionized, and follows spiral paths along magnetic field lines. The flow speed, at Earth, is around 400 km/s. The shock, at Earth, is some 17 km thick, and located 90,000 km sunward. Bow shocks exist on planets with a magnetic field, and have been observed in other star systems.

Near Earth Objects

A NEO is a solar system object whose closest approach to the Sun is 1.3 AU, and that comes in close proximity to the Earth There are 14,000 known asteroids in this category, 100 comets, solar orbiting spacecraft, and meteoroids. All these have the potential of striking the Earth. They are closely tracked from the ground, by NASA's Planetary Defense Coordination Office. A joint US/ESA project called Spaceguard is tracking NEO's larger than 30 meters. Three NEO's have been visited by spacecraft.

There more than 15,000 NEO's. Chunks of rock, hanging around Earth.. If these enter the atmosphere, they heat up and burn. Sometimes, enough is left to hit the ground. The easiest place to find meteors (as we call asteroids that hit the ground) is Antarctica, where they stand out against the snow and ice. A lot of the Antarctic meteors come from Mars, as one of my old professors proved.

Mars

Mars, and its two tiny moons and seven Trojans has got some infrastructure in place – a communications relay and a weather satellite. There are multiple Rovers and landers on the surface, including a helicopter.

The Viking program was a pair of spacecraft sent to Mars in 1975. Each spacecraft consisted of an orbiter, and a lander. A major target now is a Mars sample return mission.

The Asteroid Belt

There are millions of asteroids, mostly in the inner solar system. The main asteroid belt is between Mars and Jupiter. Each may be unique, and some may provide needed raw materials for Earth's use. There are three main classifications: carbon-rich, stony, and metallic.

The physical composition of asteroids is varied and poorly understood. Ceres, the Dwarf planet in the Asteroid belt, appears to be composed of a rocky core covered by an icy mantle, whereas Vesta may have a nickel-iron core. Hygiea appears to have a uniformly primitive composition of carbonaceous chondrite. Many of the smaller asteroids are piles of rubble held together loosely by gravity. Some have moons themselves, or are co-orbiting binary asteroids. The bottom line is, asteroids are numerous and diverse.

It has been suggested that asteroids might be used as a source of materials that may be rare or exhausted on Earth (asteroid mining), or for constructing space habitats or a refuelling stations for missions. Materials that are heavy and expensive to launch from Earth will someday be mined from asteroids and used directly for space manufacturing.

The asteroids are not uniformly distributed. In the asteroid belt, the Kirkwood gaps are relatively empty spots. This is caused by orbital resonance of the asteroids with Jupiter.

Jupiter

Jupiter has 79 known moons, and perhaps 1 million Trojans of 1 kilometer or larger. These tend to congregate at the L4 and L5 Lagrange points. The largest has a diameter of several hundred kilometers. The International Astronomical Union just announced as this book was being updated the discovery of 12 previously unknown moons of Jupiter, by an observatory high in the Andes in Chile. Only one has been named so far, Valetudo, a great-granddaughter of Jupiter The one way light time for Jupiter is 33-53 minutes.

Saturn

Saturn and it's 62 known moons has a one-way light time around 1.4 hours. Saturn has been visited by spacecraft four times. The first was a flyby by Pioneer-10 in 1979.

This showed the temperature of the planet was 250 degrees K. Voyager-1 visited in 1980. It conducted a close flyby of the moon Titan to study its atmosphere. It is, unfortunately, opaque in visible light. We do know it rains methane. Voyager-2 swung by a year later, and data showed changes in the rings since its sister mission visited the year before. Temperature and pressure profiles of the atmosphere were gathered. Saturn's temperature was measured at 70 degrees above absolute zero at the top of the clouds, and -130 C near the surface. The flybys discovered additional moons, and small gaps in the rings.

Uranus

Saturn and it's 62 known moons has a one-way light time around 1.4 hours. Saturn has been visited by spacecraft four times. The first was a flyby by Pioneer-10 in 1979. This showed the temperature of the planet was 250 degrees K. Voyager-1 visited in 1980. It conducted a close flyby of the moon Titan to study its atmosphere. It is, unfortunately, opaque in visible light. We do know it rains methane. Voyager-2 swung by a year later, and data showed changes in the rings since its sister mission visited the year before. Temperature and pressure profiles of the atmosphere were gathered. Saturn's temperature was measured at 70 degrees above absolute zero at the top of the clouds, and -130 C near the surface. The flybys discovered additional moons, and small gaps in the rings.

Neptune

Neptune has 14 known moons, and 18 known Trojans. It's one-way light time is around 4.3 hours. Neptune has also been visited by Voyager-2 in 1989. It discovered six new moons. That is the extent of close-up observations of the planet. Neptune has rings, like Jupiter and Saturn, and a great dark spot. It's moon Triton has geysers and polar caps. Triton has an interesting retrograde orbit – it goes in a different direction than the other moons. Triton's surface is mostly frozen nitrogen, and is geologically active. It is speculated that Triton has a subterranean ocean. The moon Ptoteus is an ellipsoid, not a sphere.

Pluto and beyond

Pluto was downgraded from a planet to a Kuiper Belt object. The *New Horizons* mission to Pluto and the Kuiper Belt began in January of 2006, and reached the vicinity of Pluto in July 2015. It conducted a 6-month survey of Pluto, and went out farther into the Kuiper belt, on an year extended mission, which is ongoing at this writing. It has been working fo 22 years. The spacecraft was developed for NASA by the Johns Hopkins University Applied Physics Lab in Laurel, MD.

Pluto had one known moon, Charon, before the New Horizons Team members, using Hubble Space Telescope data, discovered four more, Nix, Hydra, Styx, and Kerebos.

Kuiper Belt

The Kuiper Belt extends from the orbit of Neptune out approximately 50 AU. There are three known dwarf planets, the former planet Pluto and two others. Over 100,000 units are speculated to exist. Neptune has a major gravitational influence over the Kuiper belt objects. Not much is known about the belt and its objects, since astronomers have had to rely on ground based observation. The New Horizons mission is proceeding out through the Kuiper belt, and will tell us what it sees.

Beyond the Kuiper belt is the Scattered disc, extending beyond 100 AU. This is a sparsely populated region of the solar system. There is no knowledge of how many scattered disk objects (SDO's) exist. The closest is at around 30 AU. The belt extends above and below the ecliptic plane, in a torus configuration.

Beyond this is the Oort Cloud, extending out 5,000 – 100,000 AU. There is a disk-shaped inner cloud, and a spherical outer cloud. And you thought space was empty.

Dwarf Planets

The dwarf planets of our solar system include Ceres. Orcus, Pluto, Salacia, Varuna, Haumea, Quaoar, Makemake, 2007 OR10, Eris, and Sedna. These smaller objects did not make the size cut to be a real planet. These all orbit the Sun. Ceres is located in the asteroid belt. Orcus is a trans-Neptunian object, Salacia, Haumea, Quaoar, Makemake, and Varuna are Kuiper Belt objects. Eris is the largest of the dwarf planets, having its own moon. Sedna is beyond the Kuiper belt. It's orbital period or year is 11,400 Earth years. It's in a highly elongated orbit, probably due to Neptune's gravity. Generally, a dwarf planet does not have enough gravity to clear its orbit of other material. Not all dwarf planets have yet been discovered or observed. There may be 10's of thousands.

Comets

There are some 5,253 known comets. The Deep Impact mission returned images of the surface of comet Borrelly in 2001. The surface was hot (26-70C), dry, and dark. In July of 2005, the same mission sent a probe into Comet Tempel 1. It created a crater, allowing imaging of subsurface material. Water ice was detected. Comet Borrely has a coma, or tail, which proved to be vaporized subsurface water ice. Deep Impact went on to complete a flyby of Comet Hartley-2 in 2010.

That seems like a lot of things to look at through a telescope, or send missions to explore. But, that's just our solar system, with 1 star. There are a lot of stars, some known to have their own planetary systems.

The stars

The stars we see from Earth seem to be arranged in patterns, forming what the ancients saw as constellations. In reality, from out point of view, there are patterns, but those stars are at widely ranging distances from Earth. We can determine distances from a physics phenomena called the red shift. First, we determine if the light from the star is shifted to a higher or lower frequency. How do we know what it started out to be? The spectral bands of common elements is observed. All of the elements' spectral lines will be shifted the same amount. Comparing that to a "normal" spectrum will give the shift. If there is a red shift, there is an increase in wavelength, and the observed object is moving away from us. Similarly, a blue shift is a decrease in wavelength, and the object is moving toward us. The amount of the shift gives the velocity. The amount of shift is affected by relativistic effects at velocities approaching the speed of light.

A special class of pulsating stars is called the Cepheids. They are noticeable by changes in brightness, and usually exhibits stable behavior. There is a direct correlation between the luminosity of the star, and its period. Knowing that, scientists can use those stars as measures of distance. The first was discovered in 1784, but the behavior was not understood until around 1908. The pulsations can be on the order of days or months.

A nova is a transient astronomical event involving a bright new star that fades over time.

The Main Sequence of Stars is a chart of color versus brightness. The chart is referred to as the Hertzsprung-Russell diagram. Most known stars fit within the diagram.

A binary star is a star system, with the two stars orbiting a common point. If there's more than two, it is just referred to as a star system.

A Neutron star is the collapsed core of a very large star. The original star had to have a mass of around 10-30 times greater than our Sun. They are small and dense, with a radius of around 10 kilometers, and weighing about 2 solar masses. They result from a supernova explosion of the original star, followed by gravitational collapse which compresses what's left to the density of atomic nuclei, with no spaces. They have large magnetic fields, and rotate very quickly.

A supernova is an end-of-life explosion of a star. Observed from Earth, there is a brilliant new bright light in the sky, that gradually fades. There have been three supernova events observed in the Milky Way in the last thousand years.

A dwarf star is smaller than our Sun. They come in white, black, brown, red, yellow, orange and possibly blue. These stars are not dense enough to collapse into a neutron star at end-of-life. A black dwarf no longer emits visible light. A blue dwarf has not been observed, but is postulated to exist.

A black hole is an object that has a gravitational field that is too strong for even light to escape. It is postulated to warp spacetime itself. The boundary around the black hole is known as the event horizon. Black holes may be the result of the collapse of a very massive star. These are usually found in the galactic core. They create gravitational waves, which have just recently been observed. The concept of a Black Hole, which is essentially a point mass, was first suggested in 1784. There is a supermassive black hole at the center of the Milky Way.

A Galaxy, like our own Milky Way, is a collection of stars gas, dust, and dark matter, gravitationally bound together. The objects orbit the galaxy's center of mass, and can be of an elliptical, spiral, or irregular shape. Between the stars and other objects in a galaxy is a gas, with a density around 1 atom per cubic meter. There are estimated to be around 10^{11} - 10^{12} galaxy's in our observable universe, each having 10^8 to 10^{14} stars. It's a big place.

The Magellanic Clouds are made up of two dwarf galaxies in the southern sky, The cloud orbits our Milky Way Galaxy. These phenomena were noted by the Khoisan People of Southern Africa. The early Polynesian peoples used them as navigation aids. There is a large cloud, and a small cloud.

A Nebula is an interstellar cloud of dust and gases, spanning millions of light years. The most prominent one, from our viewpoint, is the Orion Nebula, visible with the naked eye. Stars are formed in nebula, from the concentration of matter. Ptolemy the Greek observed a nebula around the year 150. The Andromeda Nebula was observed by a Muslim astronomer in the 10th century. He named it, "Little Cloud." Later, Edwin Hubble identified it as a different galaxy than ours.

A Pulsar is a magnetized, rotating neutron star. Like a lighthouse, it emits a beam of electromagnetic radiation in one direction. Observed from Earth, it looks like it is pulsing. The stars have very regular rotation rates, and thus the pulses appear at very precise intervals. They can be used to time other events. They were discovered in 1967.

Star clusters are groups of stars. A Global cluster has thousands of old stars, an open cluster is more loosely configured, and consists of younger stars. A cluster tend to be spherical, and the constituent stars tend to be the same age. The Pleiades are an example of an open star cluster.

From somewhere else

Not everything in our solar system is from around here. Generally, our solar system bodies orbit the Sun in a disk called the ecliptic plane. Comets that are not necessarily in orbit around our Sun can take a path that are highly inclined to that plane. As this book was being written, the first asteroid at a very high angle with respect to the ecliptic was observed. That means it did not originate in our solar system, but came from some where else in our galaxy, or beyond. There is no particular name for this class of objects, but the title "Exeroid" has been suggested. This has to be cleared with the International Astronomical Union. The object was observed by the telescope on the Hawaiian mountain of Hakeakala. It is the first time an interstellar object has been observed. It was named Oumuamua, in Hawaiian, "messenger from afar, arriving first." It came from the direction of the constellation Lyra. The messenger was first spotted on October 19, 2017. It is currently out of the view of Earth-based telescope. There is talk of launching a mission to track it down, by the Institute for Interstellar Studys.

Exoplanets

Exoplanets are planets orbiting another star than our own Sun. Although it is difficult to see them through a telescope, we can define them by their gravitational effects on their primary (star). Now, with the JWST, we have expanded our search zone by a huge factor.

A Galaxy, like our own Milky Way, is a collection of stars, gas, dust, and dark matter, gravitationaly bound together. The objects orbit the galaxy's center of mass, in an elliptical, spiral, or irregular shape. Between the stars and other objects in a galaxy is a gas, with a density around 1 atom per cubic meter. There are estimated to be around 10^{11} - 10^{12} galaxy's in our observable universe, each having 10^8 to 10^{14} stars.

At the moment, we know of around 4,00 planets orbiting other stars. We don't know how many are in the habitable zone. Even the definition of "habitable" is not well defined, and goes according to "life as we know it." Exoplanets can be in any orbit – they can rotate in the same direction they are moving in orbit, or the opposite way (retrograde). There are also rogue planets, that do not orbit a star. It is thought that there may be a billion of these in the Milky Way galaxy alone.

Just recently, the European Organisation for Astronomical Research'ss Very Large Telescope in Chile took a picture of the brown dwarf 2M1207. It also caught a Jupiter-sized planet, termed 2M1207b.

We are just beginning to understand planet formation. We think we understand star formation.

More than one exo-planet can orbit a star, giving us exo-solar systems. In addition, an exo-moon has been discovered orbiting a known exoplanet. Recently discovered was something that had never been seen before, and needed a name. It was a small body, orbiting the moon of a known exoplanet. The astronomers decided to call it a *moonmoon*. Beyond that, seven known exoplanets orbit a binary star system. There can be two planets that orbit each other, without a star.

Some observed exoplanets are rocky, like Earth and the inner planets. Some as gas giants, like Jupiter and Saturn. Some are icy giants, like Neptune and Uranus.

There is an upper limit to the size of a planet. If it is more than about 12 Jupiter masses, it will evolve into a star, not a planet. The gravity at that level is enough to crush matter, and start nuclear fusion. This rather small star is called a brown dwarf.

There are many planet classifications, including Giant, Super Earth, Super-Jupiter, Sub-Earth, Mini-Neptune, a Planetar (brown dwarf or smaller), Planemo (not quite a star), and Mesoplanet, smaller than Mercury, but larger than Ceres.

In addition there are the Hot Jupiters, unusually close to their star. Closer, in fact than our planet Mercury. They have an orbital period of 10 days or less. The hot Jupiter 51 Pegasi b was found in 1995.

Super Puffs are Jupiter-sized exo-planets that are about 10 times lighter. There is one close to the star WASP-

107P in Virgo. It orbits its star in 5.7 days.

There are so many exoplanets known that I will only discuss a few. The exoplanet takes on the name of its parent star, with a number representing the order in which it was discovered.

The largest exo-planet observed is HR2562, with a mass about 30 times that of Jupiter, On the other side of the spectrum, the smallest observed is about twice the size of Earth's moon. The nearest, discussed below, orbits our neighbor star Proxima Centauri, a bit more than 4 light years away.
And there are exo-solar systems, with collections of exoplanets, with their exo-moons. It is a strange Universe out there.....

Space Weather

The Sun controls the weather in our solar system. Large solar flares can release up to 10^{25} joules of Energy. The Sun releases electrons, stripped from their atoms, the resulting ions, and intact atoms from the corona. It also releases radio waves, which, traveling at the speed of light, reach Earth 8 minutes later. The particles travel at sub-light speeds. Bright aurora's in the polar regions will be created. Stellar flares have also been observed on other stars.

When the solar wind reaches the Earth's magnetic field, it interacts with it, creating a Geomagnetic storm. There can also be proton storms from the Sun. One effect is that the upper atmosphere is heated to tens of millions of degrees Kelvin. Luckily, it is near-vacuum. Increased electromagnetic radiation from radio to Gamma Rays are also observed.

NASA's *Wind* spacecraft was launched in 1994 to study radio waves and plasma in the solar wind. It continues to operate at this writing, from the Earth-Sun L1 point

As a side note, the out-streaming solar wind can be used with solar sails to catch a ride, in the same way a sailboat is moved by the wind on Earth. It is even possible to tack in towards the Sun. Solar sails are lightweight, and need to be reflective.

The Sun has an 11 year cycle, ranging between being fairly quiet, to very stormy. Solar flares are not completely understood, and there is no good model for their prediction. Generally, we have a 2-hour window between detection, to when the storm hits.

Space weather has two observational goals – monitoring the Sun, to see what it is generating, and observing the effects on the planets and moons of our solar system.

Astronomy Missions

This section has overviews of the Astronomy missions to date, with some discussion of future missions. Unless a satellite is providing services such as Direct-TV or GPS, it is either looking up, or looking down. If it's looking down, it is a weather or climate mission. If it's looking up, its an astronomy mission. That's how we rocket engineers characterize things. We will discuss missions here that look up.

Airborne Observatory

This section discusses platforms with a telescope, including planes, and airships. Infrared observing missions can fly above the blocking water vapor in the atmosphere. The first balloon borne astronomy mission was Stratoscope I in 1957. The Boomerang (Balloon Observations Of Millimetric Extragalactic Radiation And Geophysics) project looked at the cosmic background microwave radiation. It flew above 125,000 feet. First flight was in 1997. In 1998 and 2003, it ascended from McMurdo Station in Antarctica. The mission duration on this flight was 2 weeks.

NASA's Galileo Airborne Observatory was hosted on a Convair 990 aircraft. It was first used in 1965, but was unfortunately destroyed n a mid-air collision in 1973.

NASA's Kuiper Airborne Observatory used the large Lockheed C-141 Starlifter. It operated from 1974 to 1995.

The Joint NASA-DLR (Stratospheric Observatory for Infrared Astronomy (SOFIA) has been in use since 2010. It uses a Boeing 747SP with a 2.7 meter telescope. The aircraft is based at the Armstrong Flight Research Center in Pasadena, CA. The telescope is specially mounted to isolated it from movements of the plane.

NASA's Astronomy satellites

This section discusses the various space telescopes projects by NASA. Most control centers for the unmanned astronomy missions in Earth orbit are at the

Goddard Space Flight Center, with Science Data Processing centers hosted there, or at Universities. The Jet Propulsion Lab in Pasadena, California, handles missions that go out beyond Earth orbit, and visit other planets. They have a couple of spacecraft that have left our solar system entirely. Both facilities have visitor's centers, if you want to see the hardware. I would also recommend the Smithsonian's Air & Space Museum in Washington, D. C. and its sister facility at Dulles Airport in Virginia.

Some astronomy is done from our big "NASA Center" in orbit, the International Space Station. This provides a convenient place to attach an observing payload, and return it to Earth later.

In addition, some observing instruments went as "up-and-back" payloads on the Shuttle when it was operational. This limited the mission time, but you got your instrument back to possibly fly again. The missions usually involved the Spacelab module in the Shuttle bay, and associated pallets, with pointing systems, controlled from the Shuttle itself, or the ground. One such payload was Astro-1, which included three telescopes. After many delays, it flew on Shuttle Columbia in December of 1999. The Astro-2 payload went in 1993 on Endeavor. It had a Ultraviolet Imaging Telescope, the WUPPe (see glossary), and an Ultraviolet telescope from Johns Hopkins University.

Astronomy missions are usually designed to observe in

specific spectral bands such as infrared or ultraviolet. In general, different specialists are in charge of these missions. I know of one case where a Gamma Ray scientist and an X-ray astronomy were having coffee at the cafeteria, and discussing work. They found that a gamma ray burst had occurred at the same time and in the same area as an x-ray burst. Serendipitous, and something new, cross-disciplinary was learned.

HETE

The High Energy Transient Explorer was a NASA joint project with France and Japan. It was built to observe gamma ray bursts, and had UV, X-ray, and gamma ray detectors. The HETE-1 mission ended when the satellite failed to separate from its launch vehicle attach fitting and co-payload in 1996. The replacement HETE-2 was launched in 2000. The mission is still operational, but was ended by aging batteries.

See, http://space.mit.edu/HETE/mission_status.html

Swift

The Swift Gamma Ray Burst Explorer was launched in 2004. It's main objective was to determine the origin of gamma ray bursts. It has to date taken over a million images, and is still operational. The control center is on the campus of the Pennsylvania State University. A gamma ray burst can be accompanied with an afterglow in the UV and optical spectrum areas.

There are three primary instruments. The Burst Alert

Telescope (BAT) detects bursts, and computes the coordinates. The computation takes 15 seconds, and the coordinates are downlinked to the control center, for distribution to ground-based observatories. The X-ray telescope performs spectral analysis on the afterglow. This helps to more accurately determine the location of the burst. The afterglow can last for days to weeks. The ultraviolet and optical telescope looks for an optical afterglow after a burst.

This project was renamed the Neil Gehrels Swift Observatory, after Dr. Neil Gehrels of NASA/GSFC, Principal Investigator, who passed away in 2017. He was my friend, and the "go-to" guy for Gamma Ray bursts.

A planned successor to Wwift is the Space Variable Objects Monitor (SVMOS), being developed by China and France, for launch in 2023.

Fermi

The Fermi Gamma Ray Space Telescope, was originally termed Glast (Gamma-ray Large Area Space Telescope). It was a follow-on to the Compton mission, and was launched in 2008. It was a joint effort between NASA, DOE, France, Germany, Italy, Sweden, and Japan. It is complemented with observations from the HETE and Swift missions. The main observing instrument is the Large Area Telescope. It has a dedicated Gamma Ray Burst Monitor. At the moment, it is the most sensitive gamma ray instrument in orbit. The mission was designed for a 5-year operational lifetime, with a goal of

10 years. It made the 10-year goal, and is continuing to return useful data. One of the rotating solar arrays is stuck, causing operating constraints in power, and thermal issues.

Uhuru

Uhura was the first satellite specifically designed for X-ray astronomy. Uhuru means *Freedom* in Swahili. The mission was launched from the San Marco platform near Mombasa, Kenya. It was officially NASA's Explorer 42. It was a scanning instrument, with a spin period of about 12 minutes. It did an all-sky survey in the X-ray spectrum in the 2-20 kev range. It used proportional counters sensitive to photons in the X-ray spectrum. It discovered several new X-ray sources. After returning data for two years, the mission ended in 1973.

Small Astronomy Satellite, SAS-2

The Small Astronomy Satellite-2 (number 48 in the Explorer series) was launched in 1972. It carried a gamma ray telescope, and launched from a platform in the ocean off Kenya, into low Earth orbit. It operated from November 1972 to July 1973, when a failure in a power supply closed down the mission. It's major claim to fame is discovering the Pulsar Geminga.

The mission objective of the satellite was to measure the spatial and energy distribution of primary galactic and extra-galactic gamma radiation, with energies in the range between 20 and 300 MeV.

OAO

The first Orbiting Astronautical Observatory satellite was lost in 1966 when a power supplied failed 3 days into the mission. The replacement OAO-2 was launched in 1968. Nicknamed *Stargazer*, it observed in the Ultraviolet, using 11 telescopes. What was to be OAO-3, termed OAO-B at launch, never made it to orbit, due to a separation issue with the upper stage. The OAO-3 mission was launched in 1972, as a collaboration with the Science Research Council in the UK. It operated until 1981. It was based on NASA's Multi-Mission Modular Spacecraft bus, and built by Grumman Aerospace in New York. It used the NASA Standard Spacecraft Computer. It was operated from the GSFC. At some point, you will get tired of me saying, "I worked on that mission," but I did. OAO-3, named *Copernicus*, was a 1972 mission. A collaborative effort with the UK, it was the most successful in the series. It had an X-ray detector, and a UV telescope.

HEAO

There were three spacecraft in the High Energy Astronomy Observatory series. Only number two acquired a name, the Einstein observatory. The HEAO program operated from NASA's MSFC, from the late 1970's through the 1980's. These spacecraft observed in the x-ray and Gamma-ray spectrum. They were placed into near-circular orbits, with an altitude around 500 km.

HEAO-1's mission was a sky-survey, which it conducted from 1977 through 1979. It had four different instruments, provided by GSFC, NRL, the Smithsonian Astrophysical Observatory, Harvard College observatory, and the University of California at San Diego.

The mission resulted in the compilation of the HEAO A-1 X-ray Source Catalog, with 842 identified sources.

HEAO-2 was launched in 1979, and operated until 1981. It had a single large focusing X-ray telescope. At the focal plane were four instruments, a high resolution imaging camera, an imaging proportional counter, a solid state spectrometer, and a Bragg Focal Plane Crystal Spectrometer. It was called the Dinstein Observatory. Contact was lost in April of 1981.

HEAO-3 went into Earth orbit in 1979, and reentered in 1981. All three spacecraft were built by TRW Systems. HEAO-3 had three instruments, a cryogenic gamma-ray spectrometer, and two cosmic ray instruments. One of these measured the composition of isotopes, from beryllium to iron. The other was a heavy nuclei experiment, to help characterize cosmic ray sources. The mission ended in 1981.

IRAS

The Infrared Astronomical Satellite has the distinction of being the first mission to perform an all-sky survey in the infrared spectrum. This was a joint NASA project with the UK and the Netherlands. I was launched in January of 1983, and de-activated in November of that year. Infrared instruments are limited by their cooling systems.

In IRAS' case, this was liquid helium, which was depleted in 10 months.

One of its major findings was strange infrared signatures around some stars. Hubble was tasked to examine these, and discovered planetary disks around those stars IRAS also discovered three asteroids and six comets.

NuSTAR

The Nuclear Spectroscopic Telescope Array (NuSTAR) observed in the X-ray spectrum. It was part of NASA's Small Explorer program, as Explorer 93. It launched in 2012, and is operating as of this writing. It has exceeded its planned 2-year lifetime, and is still operating. The Earth's atmosphere is opaque to X-rays, and all we know about X-ray sources in Space comes from satellite observations. NuSTAR produced much new data on the characteristics of the sources of x-rays. It also conducted a deep survey for black holes.

RXTE

The Rossi X-ray timing explorer (Explorer 69) was a NASA mission, launched in 1995. It carried an all-sky monitor, a proportional counter array and the High-energy X-ray timing experiment. The Mission was named for the physicist Bruno Rossi. It observed x-rays originating from black holes, neutron stars, x-ray pulsars, and x-ray bursts. It had three instruments, an all-sky monitor, a proportional counter array, and a high energy x-ray timing experiment. It operated until 2012, and has since re-entered the atmosphere. It's major claim to fame was supplying the proof that the diffuse background X-ray glow in the galaxy comes from previously undetected white dwarf stars and from other stars' coronas.

WISE

The Wide Field Infrared Survey Explorer launched in 2009. It operated until 2011, when it was put into hibernation mode, but was reactivated in 2013. It's accomplishments include the discovery of thousands of minor planets, and many star clusters. It was the first to see Earth's Trojan, named 210 TK7. WISE was built by Ball Aerospace.

WISE performed all-sky surveys in the 3.4 to 22 um wavelengths. After its cryogenic coolant was depleted, a mission extension was conducted to search for near-Earth objects. WISE's data was released into the public domain in 2012, and can be found at http://irsa.ipac.caltech.edu/frontpage/. WISE took some 1.5 million images over its lifetime, each with a 6-arc-second resolution. It observed in four bands, 3.4, 4.6, 12, and 22 microns.

It provided observations of over 33,000 new asteroids, and 154,000 other solar system objects, 290 near-Earth asteroids and comets. Among its accomplishments if the discovery of the most luminous known galaxy in the Universe, known as WISE J224607.57-052635.0 .

The NEOWISE program was a 1-month extension to the mission, in October, 2010. This was successful, so the mission was extended another 3 months. Only two of the four sensors can work without cyrogen. In this phase, twenty new comets were discovered. Up to this point, some 158,000 minor planets were characterized, with 35,000 new objects. It was further extended to 2023.

They spacecraft was put into hybernation in February of 2011. Status was checked in September of 2012, and everything looked good. The coolant was depleted, tho. A year later, NASA decided to recommission the mission to search for NEO's. This was just after the Chelyabinsk meteor escaped detection, and exploded over Russia. Wise came out of hibernation in September of 2013. The telescope was pointed into deep space for a while, to lower its temperature to -200 degrees C.

The new operation was titled Neo-Wise. As of May 2018, the mission had discovered more than 250 Near Earth asteroids, nearly 50 potentially hazardous objects, and 28 new comets. Just in the period 2013-2017 it imaged 26,000 previously known asteroids and comets, and 416 new space objects. It was in extended mission until 2021.

COBE

The Cosmic Background Explorer launched in 1989, and operated until 1993. It was designated as Explorer-66, and incorporated technology from the IRAS mission. COBE investigated the cosmic microwave background radiation. It was able to collect information that supported the Big Bang Theory. Several investigators got the Nobel Prize in Physics as a result of their work on COBE-supplied data. Cobe was followed by the WMAP and Planck missions.

Cobe had three instruments, the Diffuse Infrared Background Experiment (DIRBE), the Far-InfraRed Absolute Spectrophotometer (FIRAS) and the Differential Microwave Radiometer (DMR).

The mission operated in a Sun-synchronous orbit. It eventually ran out of its superfluid helium coolant. It had been designed to operate for at least 6 months. It added ten new galaxy's to our knowledge-base. Like most NASA missions, the Cobe data is available online.

COBE's end-of-mission was in Decembner 1993.

A follow-on missions were the Wilkinson Microwave Anisotropy Probe by NASA, and ESA's Planck mission. NASA contributed to Planck. The spacecraft observes in the Microwave and infrared spectrums. It is collecting data on the cosmic background. It has improved observations over those of Wmap. It operated at the Sun-Earth L2 point until de-activated in 2012.

AMS

There were two of the Alpha Magnetic Spectrometer spacecraft, one launched in 1988, the other in 2011. The prototype AMS-11 flew on the STS-21 mission, for proof of concept. AMS-2 was to be delivered to the ISS in 2003, but the Shuttle disaster delayed these plans. Finally, it got there in 2011 on Mission STS-134. It is mounted on the Integrated Truss Structure. It weighs 8500 kg, and required 2.5 kw of power. It does a lot of analysis on collected data, resulting in a low 2 megabit/sec downlink requirement. It catches about a thousand cosmic rays per second. The operations control center are located at CERN in Switzerland.

The AMS-2 measures antimater in cosmic rays. It recorded billions of cosmic ray events, and nearly half a million positrons. It is examining basic cosmological concepts such as anti-matter, dark matter, and the existence of classes of quarks known as strangelets. The first scholarly paper resulting from the observations was published in 2013.

AMS-2 was the result of work by over 500 scientists from 56 institutions around the world, and was called the most sophisticated particle detector sent into space. Before launch, it was to be equipped with a superconducting magnet, but this was not done, due to a shortened operating life, with that approach.

WMAP

The Wilkinson Microwave Anisotropy Probe was launched in 2001, and operated until turned off in 2010. It measured temperature differences in the cosmic microwave background. This is, essentially, the left-over radiant heat from the Big Bang. The follow-on to WMAP is ESA's Planck spacecraft. WMAP helped establish a standard model of Cosmology. In this model, the Universe is dominated by dark energy.

The spacecraft had sun shields and thermal radiators to keep the low noise amplifiers to 90 degrees Kelvin or below. It was located at the L2 Lagrange point, a gravity null.

The spacecraft is now a zombie-sat in a graveyard orbit. It's legacy lives on in its data, which is publicly available. (see, resources).

Sampex

This mission was the Solar Anomalous and Magnetospheric Particle Explorer, launched in 1992 into a polar orbit. It was a collaborative effort between NASA and the Max Planck Institute for Extra-terrestrial Physics in Germany. It had a three-year planned life time, but continued to supply data through 2004.

Sampex included a telescope for large ions, an ion composition analyzer, a mass spectrometer, and a proton-electron telescope

The Sampex Operations were conducted from the Goddard Space Flight Center. At the end of its mission life, control was transferred to near-by Bowie State University, to use as a teaching tool for aspiring Satellite Controllers.

IBEX

The Interstellar Boundary Explorer was launched in 2008, with the task of mapping the boundary between our solar system and interstellar space. This is currently thought to be around 90 AU, where the solar wind slows to subsonic speed.

IBEX is in a sun-orientation, in a highly elliptical Earth orbit. It has two energetic neutral atom imagers, and electrostatic analyzer, an a particle counter It records particle counts in two bands, low and high energy.

It did an orbit adjustment in 2011 that will allow a useful lifespan of 40 years, for all the electronics hold out. So far, so good. The mission control center is located on the Orbital Sciences' Dulles facility in Virginia.

SDO

The Solar Dynamics Observatory has been on station since 2010, observing the Sun. It was part of NASA's Living with a Star Program. This sought new knowledge about the Sun-Earth System.

It is operated from GSFC, and is still returning useful data. By studying the sun's emissions, the Science Team is learning more about Space Weather, which influences the Earth.

The spacecraft images the Sun in ten different wavelengths. It has three instruments: the Helioseismic and Magnetic Imager (HMI), the Atmospheric Imaging Assembly (AIA), and the Extreme Ultraviolet Variability Experiment (EVE). HMI gives continuous full disk coverage at a high spatial resolution. AIA takes images of the solar disk in 10 different wavelengths, every 10 seconds. EVE measures the extreme ultraviolet irradiance of the Sun. The instruments take an image of the Sun every second. That is a lot of data.

Ulysses

Ulysses was a solar polar orbiter, a joint effort between NASA and ESA in 1990. It was launched in 1990, and collected data until 2009.

The mission was planned to observe the Sun at all latitudes, requiring the spacecraft to leave the plane of the ecliptic. For this, it used a gravity assist at Jupiter. The launch was delayed, as it was originally manifested on Shuttle *Challenger*.

Power was supplied by an RTG instead of solar panels, since the mission went to Jupiter first. Newer model solar arrays can operate successfully at distances out to Jupiter.

Radio waves generated by plasma were studied with long (72m) antennas. There was also an x-ray detector, two scintillators, and several magnetometers. A scintillator exhibits luminescence – it glows when struck by a particle, making a good particle detector.

The mission was extended several times, as the spacecraft was working well, and valuable data was being obtained. Besides the Sun data, valuable information was obtained at Jupiter, and during the cruise.

Ulysses showed the complexity of the the Sun's magnetic field; the fact that there was a lot more "space dust" coming into the solar system than thought, and that the Sun's magnetic field was weaker than thought.

IXPE

The Imaging X-ray Polarimetry Explorer was officially announced early in 2017. It will consist of three identical telescopes, and will be used to measure the polarization of cosmic x-rays. This provides insight into the physics of the objects that generated the x-rays, including black holes, pulsars, neutron stars, quasars, remains of supernovae, and galactic nuclei. It should be ready to fly in 2021. It is being built by Ball Aerospace in Colorado, for the Marshall Space Flight Center in Alabama. NASA is collaborating with the Italian Space Agency, and several U. S. Universities. It launched in 2021, and was built to work for two years in space. It transmits data down to a ground station in Kenya

UVC

The Far Ultraviolet Camera/Spectrograph was deployed on the Descartes Highlands of the lunar surface by the Astronauts of the Apollo-16 mission in 1972. It was tripod mounted and had a Schmidt camera with a 20 degree field of view for imaging, and could operate as a spectrograph. The astronauts removed the film cartridge, and returned it to Earth. The images are available from a NASA archive.

IUE

Here's another mission I worked on, the International

Ultraviolet Explorer. The project with joint with NASA, ESA, and the UK's Engineering Research Council. It was designed to operate for 3 years, but lasted for 18. It was shut down for financial, not technical reasons. It has the distinction of being the first satellite observatory to be operated in real-time by astronomers. There were over 100,000 observations made, from its position in geosynchronous orbit. The primary mirror was 45 cm in diameter. There were dual redundant cameras in the short and long wave areas. IUE imaged Halley's comet in 1986. In 1987, it caught a supernova exploding in the Large Magellanic Cloud. It did extensive characterization of the interstellar medium. IUE discovered the galactic Corona, a halo of hot gas around the Milky Way.

IUE used GSFC's Multi-Mission Modular Spacecraft (MMS) architecture. Although it came equipped with 6 gyros for attitude sensing, eventually five of these failed, but operational work-arounds used the fine error sensors, the remaining gyro, and the sun sensor to extend mission life. In 1995, the usage and thus cost was redistributed 1/3 NASA, 2/3 ESA. The satellite remained operational and provided data until September of 1996. It remains in geosynchronous orbit, just another zombie sat. In the future we may be able to retrieve and return of these memorable spacecraft, and alleviate the orbital debris problem at the same time. Any one at Smithsonian Air & Space interested?

All of the IUE data is available in the IUE Archives, hosted at the Space Telescope Science Institute on the campus of the Johns Hopkins University in Baltimore,

MD.

IBEX

The Interstellar Boundary Explorer was a NASA mission launched in 2008. It is in a sun-oriented Earth orbit. It has dual energetic neutral imagers, Hi and Lo. It collects particle counts in two energy regions. IBEX generally detects 600 particles per day, 500 in the high range, and 100 in the low. It has detected neutral atoms from outside out solar system, which differed from the home-grown variety.

IBEX's orbit was tweaked in 2011 by raising its perigee 30,000 km, avoiding some lunar gravitational effects. It is still delivering data.

EUVE

The Extreme Ultraviolet Explorer was launched in 1992, and operated until 2001. (Did I work on this? I can't remember.) The spacecraft observed in short wave ultraviolet, 7-76 nm. It accomplished an all-sky survey in the extreme uv. It also studied the interstellar medium in the uv. It finished up its mission, and re-entered the atmosphere in 2002. It's data can be found on the National Space Science Data Center Coordinated Archive, at NASA Goddard.

IRIS

The Interface Region Imaging Spectrograph was launched in 2013, and is still operating at this writing. It was designated as Explorer 94. It is a NASA-Ames mission, with instruments supplied by the Smithsonian Astrophysical Observatory and Lockheed Martin Solar and Astrophysics Laboratory (LMSAL). It's data is collected as high-resolution images, and fed to advanced computer models to help understand how energy and matter move from the Sun's surface to its outer atmosphere. It focuses on the Sun's chromosphere, the lowest part of the solar atmosphere. This area may be the spawning grounds of the Coronal Mass Ejections (CME's) that threaten orbiting satellites. Also, in the Sun's atmosphere, temperatures can reach a million degrees (Fahrenheit or Celsius, doesn't matter), where the surface temperature is only a few thousand degrees.

GALEX

The Galaxy Evolution Explorer was launched in 2003, and operated for 9 years. It is in a circular orbit around the Earth. After its primary mission goals were achieved in a little over two years, it continued to operate for another seven. It was placed in stand-by mode in 2012, due to funding issues. Operations were transferred to the California Institute of Technology. It did observations in the ultraviolet, gaining information on star formation. It observed hundreds of thousands of other galaxy's, and looked at the rate of star formation in each.

CHIPS

The Cosmic Hot Interstellar Spectrometer was another 2003 mission, the first of the University-class Explorers operated by the University of California, Berkeley. It is an all-sky spectroscopy mission, focusing on the diffuse background, at wavelengths of 90-260 Angstrom. This band has previously been neglected. At launch, it was expected to operate for 1 year. After 5 years, it was placed in safe-hold status.

Chips has the distinct of being the first mission to implement operations via the Internet.

FUSE

The Far Ultraviolet Spectroscopic Explorer, was a NASA Project, joint with CNES. It was operated by Johns Hopkins University, Applied Physics Lab, launched in June 1999. It observed in the far ultraviolet, 90.5-119.5 nanometers. It had an objective of characterizing the stellar processing time of deuterium left over from the Big Bang. It had a series of failures on the spacecraft bus side, and its final reaction wheel failed in 2007, and it was no longer able to maintain pointing attitude.

I worked on this, and baby-sat the spacecraft for many hours during extensive thermal/vacuum testing at GSFC.

Nancy Grace Roman Space Telescope.

This telescope Telescope will launch in 2027. It is a NASA/GSFC mission, and will observe in the infrared spectrum. It was to be named the Wide Field Infrared Survey Telescope, but that was changed to the Nancy Grace Roman Space telescope, to honor the astronomer known as the Mother of Hubble. She was Chief of Astronomy at the Goddard Space Flight Center.

It has a 2.4 meter field of view telescope, like Hubble. It has an infrared camera, and a Coronograph. That instrument is small field of view, but high contrast, in the near infrared and visible spectrum. These mission grew out of a joint NASA-DOE's Joint Dark Energy Mission. The spacecraft also assists in a census of Exo-planets. It may be able to use its chronograph for direct observation of these interesting objects.

It will have the ability to block the light from the parent star, to better study the exo-planets. It will be looking for Earth -size planets within the habitable zone. It will study the planet's atmosphere to check for transformation of the atmosphere by life. This goes far beyond what we can do today, even with JWST, where exoplanets are detected by their gravitation effects on their "Sun."

Parker Solar Probe

The Parker Solar Probe built at JHU-APL for GSFC, was launched in a mission to monitor the Sun from a near position of about 8 solar radii. Why won't it burn up? It has a highly eccentric orbit, which allows it to cool off periodically. It will spend 11 days close to the Sun, and 158 days, further out, cooling off. It also has a massive heat shield. (The old joke was, if you launch a solar mission, do it at night.) The probe will use seven gravity assists over seven years from Venus to adjust its orbit at the Sun. This is a maneuver that taps some of the planets energy to slingshot the satellite where you want it.

Parker will approach the Sun to less than 9 solar radii,

where it will be in the outer solar corona. It uses small solar panels, some of which retract behind the heat shield. It is the first in being a spacecraft named after a living scientist, a solar astrophysicist, Professor Eugene Parker at University of Chicago. He was 91 at launch.

The solar probe concept goes back to a Solar Orbiter project in the 1990. It was to use a gravity assist from Jupiter, but Parker didn't have to go that far. The Venus flybys will be completed by 2024.

The spacecraft is protected by a solar shield. It will face an intensity of 650 kilowatts per square meter. The shield, made of reinforced carbon-carbon is 4.5 inches thick. The temperature of the sun-facing side will get to 2,500 degrees F. The shield is covered by a layer of white alumina to reduce heat absorption. It was estimated that the spacecraft would be inoperable within seconds if the shield were not in place. There are two solar arrays for power. One is retracted behind the shield at 2.5 AU. Another smaller array continues to collect power, and is liquid cooled.

Instruments include direct measurements of magnetic fields, measurements of energetic protons, electrons, and heavy ions, a wide field optical imager, an analysis of solar wind constituents.

Launch and deployment went smoothly, and testing is being conducted. The first science data is expected by the end of 2018.

RHESSI

Rhessi, the Ramaty High-Energy Solar Spectroscopic Imager, was launched in 2002. It observes in some wavelengths common with the Compton and Chandra missions. It is permanently pointed toward the Sun, and observes solar flares in the hard x-ray through gamma ray range. Rhessi was the first spacecraft to measure gamma ray flashes from terrestrial thunderstorms. Luckily, its operating life spanned a full eleven year solar cycle. The mission operated until 2018, as this book was being prepared. Its results are being archived, and made available. The first science data was seen at the end of 2018.

SWAS

The Submillimeter Wave Astronomy Satellite operated from 1998-2004. It's far infrared/microwave telescope was designed by the Smithsonian Astrophysical Observatory. Its mission was to collect data on interstellar clouds, and look into the processes for stellar and planetary formation.

SWAS had been shut down in 2004, but was brought back to operation to support he Deep Impact Mission, for comet 9P/Tempel in 2005. SWAS analyzed the water ejected from the comet due to the impact. It was then shut down again, with the possibility of bringing it out of hibernation again if it is needed.

WIRE

The Wide Field Infrared Explorer was launched in 1999

into a polar orbit. It was to conduct a four-month-long infrared survey of the entire sky, looking for starburst galaxies and proto-galaxies.

Soon after launch, an on-orbit accident resulted in the rapid depletion of the solid hydrogen cyrogen, needed to cool the telescope to operating temperatures. However, the star tracker still worked, and was used for long-term photometric monitoring of bright stars. The spacecraft was also used as an on-orbit teaching tool for aspiring satellite controllers at a local college.

The spacecraft re-entered the atmosphere in 2011. The science goals of WIRE were taken up by the Wide-field Infrared Survey Explorer (WISE).

WIND

WIND is a NASA heliophysics mission to study radio waves and plasma in the solar wind. It was launched in 1994. Along the way to the L1 Lagrange point, it was used to study the magnetosphere and near-lunar environment in collaboration with the SOHO and ACE spacecraft. The spacecraft is currently operation as of this writing. It has 50 years of station-keeping fuel, but the electronics will give out long before then. It is operated from Goddard Space Flight Center. It was still operating in March of 2021.

Alexis (LANL)

The Array of Low Energy X-ray Imaging Sensors is a project of the (U. S.) Los Alamos National Laboratory. It can focus low energy x-rays or extreme ultraviolet. It was launched in 1973, as an Air Force mission. It scanned half the sky at a time, facing away from the Sun. Astronomers then looked for a visual counterpart to the extreme x-ray observations. The mission operated until 2005.

Radio Astronomy Explorer

RAE was officially Explorer 49. It was a NASA Mission in 1968, equipped with two 750 foot long antennas, and a 120 foot long dipole antenna. It observed in the frequencies between 0.2 and 9.2 MHz. Because RAE-A had significant interference from Earth, the follow-on RAE-B went to lunar orbit. It was launched in 1973, and listened in on frequencies from 25 kHz to 13 Mhz. It had a 229 meter v-antenna, a 183 m v-antenna, and a 37 meter dipole. The spacecraft was stabilized in a gravity gradient orientation. All of both missions' data is archived in the National Space Science Data Center.

POLAR

A sister mission to WIND involves the Polar satellite, studying the polar magnetosphere and aurorae. It was launched in 1996, and returned data until 2008. It conducted multi-wavelength imaging of the aurorae, measured the entry of plasma into the polar magnetosphere and the geomagnetic tail, looked at the flow of plasma to and from the ionosphere, and determined the deposition of particle energy in the ionosphere and upper atmosphere. It had eleven science instruments. These included various electromagnetic field instruments, three sensors for particles, and three imagers, for the visible, ultraviolet, and x-ray spectrum.

MSX

The midcourse space experiment was a project of the U. S. Ballistic Missile Defense Organization. It maps bright infrared sources in space. This data can be used for peaceful purposes as well. It was launched in 1996, and observed in infrared, visible, and ultraviolet light. MSX shares data with NASA, and is available in the Infrared Space Archive of NASA's Infrared Processing and Analysis Center. MSX is still operating as of this writing.

NASA's Great Observatories

NASA's Great Observatory's Program included four telescopes launched in 1990 through 2003 in Earth orbit, HST for visible light, the Compton Gamma Ray Observatory, the Chandra X-ray observatory, and the Spitzer Space Telescope, operating in the infrared part of the spectrum. A forerunner of these missions was the 1966 Orbiting Astronomical Observatory.

These space missions were named after notable astronomers. They observe in various wavelengths, and seek to discover knowledge of our own solar system, and others in our galaxy and beyond.

The Chandra Observatory was originally named AXAF; Spitzer was originally called SIRTF. Both Chandra and Spitzer were outstanding astronomers. Spitzer wrote about the advantages of a telescope above the atmosphere in 1946.

CGRO

The massive Compton Gamma Ray Observatory operated from 1991 through 2000. It's job was to detect high energy photons. It had four telescopes, for X-rays and Gamma Rays. It was delivered to orbit by the Space shuttle. At end-of-life, it was de-orbited. It was named after Arthur Compton, a Nobel Prize winner in Gamma ray physics. It's target, the X-rays and Gamma rays, do not penetrate, so astronomers were literally "in the dark" until Compton reached orbit. CGRO weighed some 37,000 pounds.

The Burst and Transient Source Experiment was searching for gamma ray bursts, which it saw about once per day, doing all-sky surveys. A burst could last for 100 seconds.

The Oriented Scintillation Spectrometer Experiment detected gamma rays in the energy range 0.05 to 10 MEV. There were four detectors. Both the source, and the background radiation were observed.

The Imaging Compton Telescope was a collaboration of universities and research institutions from the U.S. and Europe. It observed in the 0.75 t 30 MeV range, measuring both energy and arrival angle.

The Energetic Gamma Ray Experiment Telescope covered covered the high energy range from 20 MeV to 30 Gev. It completed the first all-sky survey in the range above 100 MeV.

The observatory was deliberately re-entered into the atmosphere after a failed gyroscope. It was not designed for in-orbit repair. The controlled reentry was NASA's first, and was successful.

Chandra

This mission is named after a Indian-American astrophysicist. It was originally the Advanced X-ray Astrophysics Facility (AXAF), and launched by the Shuttle in 1999. It had a highly elliptical orbit, reaching about 1/3 of the distance to the moon, to keep it out of Earth's radiation belts. The mission continues to return useful data, after exceeding its lifetime estimate of 5 years. It is now in its 19th year of observations. Chandra is the Sanskrit word for moon.

Chandra is in a 64-hour orbit around Earth, voyaging out about one third of the way to the moon. It remains above Earth's radiation belts for most of its orbit. It can observe for up to 55 hours in its 65 hour orbital period.

X-ray telescopes such as Chandra use primary mirrors coated with gold or iridium. Also, the spacecraft can do spectroscopy with transmission gratings.

Among its accomplishments are the images of X-rays in the shock wave of a supernova, X-rays in a gamma ray burst, an update of the Hubble constant, Jupiter's polar x-rays, and the list goes on and on.

In spite of having a suggested lifetime of 5 years, Chandra continues to return useful data.

Spitzer

When Spitzer was launched in 2003, it had a planned mission life of 2-5 years. It is still operating and returning useful data Lyman Spitzer was an early proponent of telescopes in space.

Spitzer is an infrared telescope, which was cooled by a supply of liquid helium. When the Spitzer mission expended it's liquid helium supply in 2009, it was operated as "Warm Spitzer," in extended mission. Only two of the instruments are operable without the coolant, but that's better than nothing.

Spitzer was placed in orbit around the sun, trailing the Earth, and drifting further outward at a slow pace. Not being in orbit around the Earth, it is not exposed to the heat output of the Earth, and better observations at lower temperatures can be achieved.

The primary mirror is 85 centimeters in diameter, and was cooled to 5.5 degrees K when the liquid helium was available. This same mirror material is being used on JWST.

The instruments include the IRAC, an infrared camera operating in two pairs of chosen wavelengths with a 256 x 256 detector array, and the infrared spectrometer observes at four wavelengths, using 128 x128 element detectors. The Multiband Imaging Photometer has three detector arrays.

One of the important observations the Spitzer made was of exoplanets. This was the first time an exoplanet was imaged. Usually, they are discovered by their gravitational effects on their parent star. The surface temperature of the exoplanet was determined, and went on to make additional contributions in the areas of formation of stars. It discovered the double helix nebula, so called because of its shape. It observed a collision between two planets orbiting a distant star, as well as a gas giant some 13,000 light years away. Spitzer and HST collaborated on the discovery.

HST

Earth-based telescopes have the problem that they have to look through the atmosphere. One way to solve that is to put them in orbit. This was suggested in 1923 by Hermann Oberth. The Large Space Telescope Project was funded in 1978. The spacecraft was named after astronomer Edwin Hubble.

The Hubble Space Telescope was placed in orbit by the Space Shuttle in 1990. It is still operating as of this writing. It is managed by the Goddard Space Flight Center, and the data goes to the Space Telescope Science Institute on the Johns Hopkins University Campus in Baltimore, MD. The idea was introduced in a paper by Lyman Spitzer in 1946, well before the "space age." Mention of the advantage of such a technique was documented as early as 1837.

The main mirror is almost 8 feet in diameter. The telescope operates in the visible spectrum, as well as near ultraviolet and near infrared, which don't make it through the atmosphere. Precursor missions include the Orbiting Solar Observatory, and the Orbiting Astronautical Observatory, both from NASA's Goddard Space Flight Center. Hubble uses a Cassegrainian reflector design, from Marshall Space Flight Center, and built by Perkin-Elmer. They subcontracted the mirror to Kodak, using Corning glass. This was coated with aluminum.

After only a few weeks in orbit, it became obvious that the mirror was flawed. Corrective optics were designed, and several Shuttle flight got the orbiting facility back to its desired accuracy. At the same time, updated equipment was installed, and a failed module was replaced.

At the time, a Final Systems test was not done on HST to save money. In retrospect, this may not have been the best choice.

Hubble has a 2.4 meter, 1-ton, mirror, with corrective optics installed in orbit by astronauts from the Space Shuttle. The updates took five Shuttle Missions, and also included change-out of some of the instruments. The onboard computers were replaced, and the original tape recorder was replaced by a solid state unit All six gyroscopes were replaced, as well as the solar panels. This illustrated the value of repairs in space.

Hubble currently downlinks about 140 gigabytes of observation data weekly. The Hubble camera is 16 megapixels.

Like most Earth-based observatories, scientists can apply to use the HST for observations. Competition is fierce for observing time. In the period 1990-1997, the director of the StScI allowed amateur astronomers access to the facility for free. Thirteen amateurs were involved.

The mission was launched in 1990, repaired and updated in space by several Shuttle Missions, and continues to return good data One of Hubble's notable achievements was the determination of the rate of expansion of the Universe.

The contributions of the HST mission to our knowledge of the Universe are impressive. Hubble observations now put the estimated age of the Universe at around 14×10^9 years.

The Hubble may last another 15-20 years. It is not currently feasible to mount another repair mission, as the shuttle fleet has been decommissioned. A robotic servicing mission may be attempted.

As this book was originally being published, Hubble was currently in safe-hold mode, due to a gyro failure. It could continue to operate, although it only has 2 working gyros for pointing and positioning. The original 3 mechanical gyros, replaced on a Shuttle mission, are failed. One of the three replacement units is balky. If the third gyro behaves itself, the mission can continue. It can operate in a reduced mode, with only two. The spacecraft is designed for servicing, but we don't have the shuttle fleet anymore. A robotic servicer is being developed at GSFC that might prove handy. It is good to have HST working with JWST as they can work together on frequency ranges.

A new Hubble re-oost is beng discussed by NASA and Space-X. It's about time, the last servicing mission was 14 years ago. It will utilize a Dragon rocket, and will be done at no cost to NASA. It is in study phase at the moment. If this works out, a reboost for other spacecraft would be considered.

JWST

The James Webb Space Telescope is the follow-on the HST. It uses updated technology and a new approach for the mirror, using 18 hexagonal segments, that are individually adjustable. The resulting mirror is 6.5 meters in diameter. JWST observes in long wavelength visible through the mid-infrared. The spacecraft will be placed in a halo orbit at the Earth-Sun L2 Lagrange point about 1.5 million miles from Earth. It has a large sun shield to block the Sun's light form interfering with the observations. The project was the top pick in the 2000 Decadal Survey. Work has been going on since 1989, primarily at the Goddard Space Flight Center, As of this writing, the telescope has been shipped out to other centers for additional testing.

JWST is named for NASA's second administrator. Unfortunately he passed in 1992, not seeing his namesake launched.

Why observe in the infrared, when it's hard? Mostly

because space dust blocks visible light.

JWST is a joint project of NASA, ESA, and the Canadian Space Agency. In all, fourteen countries were involving in construction of the spacecraft. John Mather, the Senior Project Scientist, had previously received the 2006 Nobel Prize in Physics.

Partially to avoid the HST problem with the main mirror, the JWST has a large set of small, adjustable mirrors, individually controllable with 6 actuators each. They are gold-coated beryllium. Since it observes in the infrared part of the spectrum, the detectors have to be cooled to single digits above absolute zero. A large sun shield is used to shadow the spacecraft from the Sun, Earth, and moon. The solar wind will push the sun shade and thus the spacecraft around. But, JWST has a trim tab to counter this.

The project was a concept in the mid-1990's and construction was completed in November of 2016. It was originally called the Next Generation Space Telescope (NGST), but was named after James Webb, the second NASA administrator.

Where Hubble has a 2.4 meter single-piece mirror, JWST has an 18 segment mirror for a combined size of 6.5 meters. These segments are folded for launch, and deploy in orbit. It has curved secondary and tertiary mirrors. Instruments include a near infrared camera, a near infrared spectrograph, a mid-infrared instrument, and a near infrared imager with a slit-less spectrograph. The telescope is expected to see extra-solar planets (planets around other stars). The telescope masses 6.5 tons. It was launched on an Ariane vehicle. It uses NASA's Deep Space Network for data transmission.

The instrument set includes a near-infrared camera (NIRCam), and spectrograph (NIRSpec). These have 32 megapixels. It uses a new type of gyro, the hemispherical resonance gyro, for attitude sensing. With essentially no moving parts, this type of gyro should last a lot longer. The spacecraft has about 60 gigabytes of onboard storage, and uses lithium-ion batteries. Onboard, it runs Java scripts for operations.

It is not designed to be serviceable. It has a projected lifetime of 5 years, with a goal of twice that. Originally estimated to cost a half-billion dollars in 1997, the current cost looks in the vicinity of 8.8 billion dollars. Actually, the spacecraft will include a docking ring for an Orion manned spacecraft (which has yet to fly), so in theory, it might be serviceable.

One of the more interesting tasks for the observatory will be to assist in the search and characterization of exo-planets

It will be operated from the same facility that Hubble is, the Space Telescope Science Institute, on the campus of the Johns Hopkins University in Baltimore, MD.

Delivery of the spacecraft for Integration and Test at Northrop Grumman was delayed into 2019, with launch planned for 2021. There was a problem with the deployment test of the sunshield. Nominal mission life is ten years. One limitation will be the station-keeping propellant supply. The 6200 kg spacecraft went on an Ariane-6 vehicle. HST is still operational when JWST made it to orbit, and the two can be used together. HST observes in visible light, ultraviolet, and infrared. JWST observes in the ultraviolet. The ESA telescope Herschel will be in the vicinity of L2 when JWST arrives.

I was able to visit JWST several times, when it was in Goddard's massive clean room, with student tours. Some of the components were tested in Goddard's Space Environmental Simulator (thermal-vacuum chamber). The entire telescope wouldn't fit, though. Later in the program, JWST was taken to the large chamber at Johnson Space Center, the largest in the world, developed for the Apollo program.

The Nancy Grace Roman Space Telescope is a follow-on the JWST, and is in development as of this writing. It has a preliminary launch date in 2027.

Other Nation's Space Telescope Projects

Although the United States dominants Space Telescope projects, there are other Nations that do participate, and several country's have their own. ISRO launched their Astroset telescope in 2015, and there is planning for a second mission. It is India's 1500 kg multi-wave length mission. It does observations in the visible, infrared, and ultraviolet simultaneously.

Japan had the Calorimetric Electron Telescope in 2015, looking for energetic electrons and gamma rays.

Russia's Roscosmos had the eRosita x-ray telescope. The instrument was built at the Max Planck Institute in Germany. It was launched in 2019. The carrier spacecraft is called tr-RG. It can do an all-sky survey in the medium energy x-ray range.

The Russian Salyut-1 station had a Ultraviolet telescope in 1971.

Xuntian is The Chinese Survey Space Telescope, planned for launch in 2023.

Some Planned Missions

This section discusses in-work and planned missions for Space Astronomy. The more we know, the more we see how little we really know about the Universe.

What missions get approved is determined by the Astronomy and Astrophysics Decadal Survey. This is conducted every ten years by the National Research Council of the National Academy of Sciences in the United States. This involves determining the current state-of-the-art, defines priorities, and makes recommendations to funding agencies.

The first effort was "Ground-Based Astronomy: A Ten-Year Program, 1964." This covered optical telescopes. The second report in 1972, "Astronomy and Astrophysics for the 1970s," covered both ground-based and satellite missions. The Third report was titled, "Astronomy and Astrophysics for the 1980s ." The fifth report was released in 1991, and titled "The Decade of Discovery in Astronomy and Astrophysics." Its recommendations resulted in the Chandra observatory, and the Spitzer Space Telescope. In 2001, the "Astronomy and Astrophysics in the New Millennium" report was released, suggested the Next Generation Space Telescope (now, James Webb Space Telescope). The "New Worlds, New Horizons in Astronomy and Astrophysics" report was released in 2010. It suggested that the focus should include, "the nature of dark energy; the structure, distribution, and evolution of exoplanetary systems; detailed examination of extreme processes including supernovae and the merger of superdense objects; and how galaxies and galaxy clusters formed from the early hot universe." This ambitious program lead to the WFIRST mission and several new ground based observatories.

The Russians placed a space-based telescope in orbit as part of the 5th module of the Mir Space Station. They are planning a follow-on in 2019, termed SPKTR-RG. This will be a joint mission, with Germany. It is to conduct a 4-year X-ray study.

There was a previous space telescope attached to the American Skylab facility, the Apollo Telescope Mount. Today the in-orbit International Space Station serves as a platform for various observing instruments, both pointing down at Earth, and further out in Space.

Observatories on the Moon

Putting an observatory on the backside of the Moon has proposed as early as 1832. Now we can make that happen with the Artemis missions. It's an airless environment, always shielded from the Earth. There would be 2 weeks of good observation time, followed by two weeks of daylight. Communications with Earth could be maintained with a relay satellite in a lunar "halo" orbit. Radio telescopes on the lunar backside have the big advantage of being shielded from the radio emissions from Earth, and can be used continuously. Lunar material can be used for structure, and very large mirrors can be constructed in the lower gravity. This can be a major project, when we get back to the moon.

International Lunar Observatory

This is a proposed project, by the private International Lunar Observatory Association, to put a small observatory at the lunar south pole. This currently has an expected launch date of 2019. The facility will be placed on Malapert Mountain, which is 5 kilometers tall. It has an uninterrupted line-of-sight with Earth.

NASA is looking at constructing a large mirror in situ from lunar materials. This would be done by spincasting a mixture of lunar soil, carbon nanotubes, and epoxy. It helps if you have some lunar soil to try this out. Well, the Johnson Space Center has some lunar soil, thanks to the Apollo missions. This manufactured with local materials approach means the facility can be much large, as it doesn't have to be sent from Earth. The mixture of epoxy and lunar dust is spun-cast, and forms a parabola. That is then covered with a thin layer of aluminum. They project that a Hubble-sized mirror would require 60kg of epoxy, 1.3 kg of nanotubes, and about a gram of aluminum. From local sources, 600 kg of dust would be needed. The coating of the mirror could be done easily in the near-vacuum of the lunar surface.

Virtual Observatories

One technique used on Earth-based observatories, both visible light and radio, is to coordinate multiple observations from multiple locations. This technique can also be implemented in space. A DARPA program called Virtual Satellite uses this technique to combine multiple observations from different points of view. Ideally, all the observing hardware would be homogeneous, but that's difficult unless specific missions with a multitude of observers are placed into orbit. A constellation of Cubesats is one possible approach. The technique has been used with the Swift and Spitzer spacecraft, working with ground based observatory's.

Search for Exoplanets

One of the more exciting missions is the search for planets of other stars than our own sun. Although there are nearly impossible to image directly, they can be observed as the pass through our line of sight with the distant star, and cause a small dip in the perceived brightness. Exo-planets are best seen from space.

A few thousand have been cataloged. Follow-on questions are, do exo-planets have exo-moons, and can they harbor life? To have the potential of life, the planets have to orbit the right type of star, at the right distance, called the Goldilocks zone (from a Fairy tale). Not too hot, not too cold, just right.

Exoplanets are the current hot topic, and dedicated missions have been launched to search for and characterize them.

Kepler

Kepler's mission was to search for Earth-sized exoplanets. It was launched in 2009, and is operated by JPL. It uses a photometer to measure the brightness of stars that will dip as an exoplanet passes in front of the star. The brightness of some 150,000 stars is monitored. It is in an Earth-trailing heliocentric orbit.

In 2013, the second of four reaction wheels on the spacecraft failed, which disrupted operations. It is not quite a fatal failure, as the planet-hunting can continue, under less than optimal conditions. It is currently focused on finding habitable, Earth-like planets around Red dwarf stars. The success count as of 2015 was 1,000 confirmed exoplanets discovered by Kepler, later updated to 1,284 Earth-sized exoplanets in the habitable zones of their Suns.

At launch, Kepler had the highest data rate of any mission so far, because of the 95 million-pixel detector. The primary mirror is 1.4 meters in diameter, manufactured by Corning, who did the Hubble mirror.

What the telescope is looking for is about a 80 parts per million decrease in brightness as the exoplanet moves in front of its star, from the telescope's point of view.

The orbiting telescope is operated and monitored by a control center on the Campus of the University of Colorado. Science data is downloaded once per month. A partial analysis of data is done onboard, and not all raw data is downloaded. The data management is located at the Space Telescope Science Institute on the Johns Hopkins University campus, in Baltimore, MD. Here the data is partially processed, and forwarded to NASA-Ames in California.

As of this writing, the spacecraft is near end-of-life, in sleep mode, using none of the remaining onboard fuel.

TESS

The Transiting Exoplanet Survey Satellite is a follow-on to the Kepler mission. It uses transit photometry to spot planets orbiting distant stars. Unfortunately, only about 1% of the star systems are properly aligned to use this technique, but it still worth the effort. It was launched in 2018. The primary mission objective is to survey the brightest stars near the Earth. At its launch, about 3,800 exo-planets were known. TESS is expected to add 20,000 more. It will do a all-sky survey on about 85% of the sphere it occupies.

TESS will pass along to JWST, when that mission is launched, targets of interest. The TESS mission is lead by MIT, with funding by Google. It was built by Orbital

Sciences, and launched on a Falcon-9. It is in full operation at this time. It uses four wide-field CCD cameras. TESS is operated by NASA, and the Smithsonian Astrophysical Observatory.

It's first sighting of an exoplanet was in September, 2018, as this book was in preparation.

Waypoint-1

Waypoint-1 is a commercial space telescope project by a California company. It is being developed to provide services for a price. You will be able to buy observing time, and the data can be downloaded to your phone. Sort of a, rent-an-orbiting-telescope. Keep an eye on this one.

Wrap-up

There are more interesting ideas than there are money and time. How do certain missions get selected?

On part is based on need. If there is a gap in out observational tools, that would weight heavily in the mission selections. Every 10 years, the Planetary Society goes through the decadal survey process. This sets the goals for the following 10 years, based on available funds, scientific need, and best guesses. If we find that first tiny plant sprouting on Mars, priorities change.

We only have to find life on other than our planet once, and the door is open. We think now we are unique in the Universe, but the Universe is very large. We still find strange life forms on Earth, like the extremeo-philes deep under the sea, feeding off volcanic vents. We're not yet looking hard enough, and can't see far enough to write off the possibility of alternate life in the Universe. This could get exciting fast.

Glossary of Terms

See also, https://en.wikipedia.org/wiki/List_of_astronomy_acronyms

AAS – American Astronomical Society.
ABRIXAS - Broadband Imaging X-ray All-Sky Survey.
AGILE-Astro-Rivelatore Gamma a Immagini Leggero (Italy)
AGN – active galactic nucleus with a relativistic jet of ionized matter.
Alexis - Array of Low Energy X-ray Imaging Sensor.
Amor asteroids – a class of near Eartha asteroids.
AMS - Alpha Magnetic Sprectometer.
Angstrom – measure of length, 0.1 nanometer.
ANS - Astronomical Netherlands Satellite.
AO – adaptive optics
APL – Applied Physics Lab of the Johns Hopkins University.
Apogee – furthest point from a primary, in an orbit.
Apollo asteroids – class of near-Earth crossing asteroids
Apsys – extreme point in an orbit, closest or furthest.
Arc-minute – 1/60 of a degree.
Arc-second – 1/3660 of a degree.
ASCA - Advanced Satellite for Cosmology and Astrophysics, Japan.
ASIC – Application specific integrated circuit.
ASC- LPI - Astro Space Center of PN Lebedev Physics

Institute, Russian Academy of Sciences.
ASIN – Amazon Standard Inventory Number.
ASIS - Advanced CCD Imaging Spectrometer (Chandra).
ASM – all sky monitor
Asteroid – a chunk of rock; a minor planet.
Asteroid belt – a disc in the solar system between Mars and Jupiter, occupied by chunks of rock in various sizes.
Astrometry – measurements of positions and movements of objects in space.
Asteroseismology – study of oscillations in stars.
Aten asteroid – Earth crossing asteroids.
Atira asteroids – orbits are entirely within Earth's orbit
AU – astronomical unit, mean distance from the Earth to the Sun, 93,000,000 miles.
AXAF - Advanced X-ray Astrophysics Facility (Chandra).
BATSE - Burst and Transient Source Experiment, (CGRO).
BBXRT – Broad Band X-ray telescope.
Big Bang – current cosmological model for the Universe.
Binary star – two stars in orbit around a common point.
Black hole – a collapsed star, compressed so dense that not even light can escape; a singularity.
Blazar – active galactic nucleus with a relativistic jet
Blue Moon – intercalary moon, 13 full moons in a year instead of 12.
Blue Shift – an apparent shift of electromagnetic radiation toward decreasing wavelength.
BRITE – (U.K.) Bright Target Explorer
CCD – Charge Coupled Device (like in your cell phone

camera)
C&DH – command and data handling.
CDR – critical design review
Centaur – a minor planet in an unstable orbit, behaving like an asteroid or comet.
Cepheid – a variable star.
CERN - Conseil européen pour la recherche nucléaire, European Organization for Nuclear Research.
Cheops – ESA, Characterizing ExOPlanets Satellite
Cluster – groups of stars.
CMB – Cosmic microwave background.
CME – Coronal Mass Ejection – burst of plasma and magnetic fields from a star's corona.
CNES - Centre national d'études spatiales. (France)
CNRS – Centre National de la Recherche Scientifique (France)
CNSA – China National Space Administration.
COBE - Cosmic Background Explorer.
Coma – Comet's tail
Comet – a solar system object consisting of ice, dust, and gas, in highly eccentric orbit.
Constellation – patterns we see in collections of stars.
COROT (French) Convection, Rotation et Transits planétaires.
Cosmic ray – high energy radiation, from outside the solar system.
COSPAR – Committee on Space Research, International Council for Science.
CSA – Canadian Space Agency.
CTP – Command and Telemetry Processor
CXE – Cosmic X-ray experiment, HEAO-1.

CXO – Compton X-ray Observatory.
Cyrogenic – at very low temperature.
Dark energy – hypothetical form of energy that explains why the Universe is expanding.
Dark Matter – existence postulated. Might account for 85% of the matter in the known universe.
DARPA – (U.S.) Defense Advanced Research Project Agency.
Deuterium – isotope of hydrogen
DLR – (German) Deutsches Zentrum für Luft- und Raumfahrt.
DMC – data management center.
DOE – (U.S.) Department of Energy.
Dwarf planet – planet below a certain size.
Dwarf star – small star, much smaller than our Sun. Comes in   white, red, blue and black variations.
EBL – Extragalactic background light.
Ecliptic – the apparent path that the Sun seems to follow, the same as the Earth's orbit.
EDU – engineering design unit.
EGRET - Energetic Gamma Ray Experiment Telescope (CGRO).
ELDO – European Launcher Development Organization, merged into ESA.
ELT – extremely large telescope.
EMR – electromagnetic radiation.
Equinox – 2 days per year when there are equal periods of daylight and darkness.
ESA – European Space Agency.
ESRO – European Space Research Organization, merged into ESA.

EU – European Union.
EUV – extreme ultraviolet, wavelengths from 10nm to 124nm.
EUVE - Extreme Ultraviolet Explorer.
ev – electron volt, unit of energy.
Exa- 10^{19}
Exeroid – proposed name for a asteroid-like body from out of outer solar system.
    ExoMoon – orbits an exo planet
    Exoplanet – planet in another solar system.
FGST – Fermi Gamma ray Space Telescope.
FSM – fine steering mirror (JWST)
FUSE - Far Ultraviolet Spectroscopic Explorer.
Galaxy – a loosely coupled collection of 10^8 to 10^{14} stars.
Galex - Galaxy Evolution Explorer.
Gamma ray – EMR from radioactive decay.
Gas giant – a large planet consisting mostly of hydrogen and helium. Jupiter and Saturn.
GEV – giga (10^9) electron volts
GLAST - Gamma Ray Large Area Space Telescope.
GN – NASA's ground network.
GOALS - Great Observatories All-sky LIRG Survey, (Spitzer, Chandra, CHIPS, HST)
Gravitational Wave – a disturbance in the fabric of space-time.
GRB – gamma ray burst
GRO – Gamma Ray Observatory.
GSFC – Goddard Space Flight Center.
GTC - Gran Telescopio Canarias
HabEx - Habitable Exoplanet Imaging Mission
Hard x-ray – energy above 5-10 kev.

HEAO – High Energy Astronomy Observatory.
HEASARC - High Energy Astrophysics Science Archive Research Center.
Helioseismology – study of oscillations in our Sun.
Heliosphere – a volume of space dominated by the Sun. In our case, out beyond Pluto.
HETE – High Energy Focusing Telescope.
Heliocentric – sun-centered.
HETE – High Energy Transient Explorer
Hexte – instrument on RXTE spacecraft – High energy x-ray timing experiment.
HRMA – High resolution Mirror Assembly (Chandra)
HST – Hubble Space Telescope.
Hubble – Space Telescope named after Edwin Hubble.
Hubble Constant – rate of expansion of the universe.
HUDF – Hubble Ultra-Deep Field.
HUT - Hopkins Ultraviolet Telescope
HXMT – (China) Hard X-ray Modulation Telescope.
IAU – International Astronomical Union.
IBEX - Interstellar Boundary Explorer.
Ice giant – large planet consisting of ices of various substances – Uranus and Neptune.
ICSU – International Council for Science.
IIA – Indian Institute of Astrophysics.
ILO – International Lunar Observatory.
Infrared cirrus -cloud-like structures in space that emit infrared light.
INTA - Instituto Nacional de Técnica Aeroespacial, Spain.
Integral - INTErnational Gamma Ray Astrophysics Laboratory

IPAC – (NASA's) Infrared Processing and Analysis Center.
IPN – Interplanetary Network – spacecraft with gamma ray burst detectors.
IRAC - Infrared Array Camera (Spitzer)
IRAS – Infrared Astronomical Satellite.
ISIM – Integrated Science Instrument Model.
IRIS - Interface Region Imaging Spectrograph.
IRS - Infrared Spectrograph (Spitzer).
IRSA - Infrared science archive.
IRT – Infrared Telescope (Shuttle mission).
ISAS – (Japan) Institute of Space and Astronautical Science.
ISBN – International Standard Book Number.
ISIM – (JWST) Integrated Science Instrument Module
ISM – interstellar medium
ISO – Infrared Space Observatory; International Standards Organization.
ISRO – Indian Space Research Organization.
ISS – International Space Station.
IXAE – (India) X-ray Astronomy Experiment.
IXPE - Imaging X-Ray Polarimetry Explorer.
JAXA – Japanese Aerospace Exploration Agency.
JDEM – Joint Dark Energy Mission (NASA, DOE)
JHU – Johns Hopkins University.
Jovian – pertaining to Jupiter.
JPL – NASA's Jet Propulsion Lab, Pasadena, California.
JWST – James Webb Space Telescope.
Kaistsat - Korea Advanced Institute of Science and Technology Satellite.
KAO – (NASA) Kuiper Airborne Observatory.

KARI – Korea Aerospace Research Institute.
Kev – kilo-electron volts, a measure of energy
KBO – Kuiper Belt objects.
KOI – Kepler Objects of Interest.
Kuiper Belt – disc in the solar system from Neptune out about 50AU.
Lagrange point – null in the gravity field.
LANL – (U. S.) Los Alamos National Laboratory.
LASP - Laboratory for Atmospheric and Space Physics (U. Colorado)
LASS - Large-Area Sky Survey, HEAO-1.
LEGRI - Low Energy Gamma-Ray Imager.
Leica – low energy ion composition analyzer
LEO – low Earth orbit.
LETGS - Low Energy Transmission Grating Spectrometer (Chandra)
LGM – little green men.
Light pollution – interference from background sources.
Light year – the distance light travels in one year. 9.5×10^{12} kilometers.
LIRG – luminous infrared galaxy.
LISA – Laser Interferometer Space Antenna.
LMSAL - Lockheed Martin Solar and Astrophysics Laboratory.
LRR - launch readiness review.
LUCI – Lunar Ultraviolet Cosmic Imager (India)
Lunation – synodic month; average period of the moon's rotation.
LUT – (China) Lunar-based ultraviolet telescope.
LUVOIR - Large UV Optical Infrared Surveyor.
LWS – Living with a Star, NASA Program.

Magellanic clouds – two dwarf galaxies in the southern sky.
Magnetar – neutron star with powerful magnetic field
Magnetotail – a long stream of charged particles, held by magnetic forces.
MAST - Mikulski Archive for Space Telescopes (named for a Maryland Senator)
Mbps – 10^6 bits per second.
Mbytes – mega (10^6) bytes.
MCC – Mission Control Center
MEV – million electron volts, a measure of energy
microlensing – light being bent by the gravity of a large object.
micron – micro meter
Milky Way – our Sun is in this Galaxy.
MIPS - Multiband Imaging Photometer (Spitzer)
MIT – Massachusetts Institute of Technology.
MMOC – Multi-Mission Operations Center, GSFC, Bldg 14.
MMS – multimission modular spacecraft
Moon – smaller astronomical body in orbit of a planet.
MSFC - Marshall Space Flight Center.
MSX – midcourse space experiment
Nanometer – 10^{-9}
NASCOM – NASA communications network.
NASDA – National Space Development Agency (Japan)
NEA – near Earth asteroid.
NEC – near Earth comet.
NASA – National Aeronautics and Space Administration.
Nebula – interstellar cloud of dust and gasses.
NeN – (NASA) near Earth network

NEO – near Earth object.
Neutron star – collapsed core of a large star. Very dense.
New Horizons – imaging mission to Pluto and beyond.
NGC – New General Catalog (of Nebulae and Clusters of Stars).
NGST – Next Generation Space Telescope – renamed after James Webb.
NIAC - NASA Institute for Advanced Concepts.
NICER – Neutron star interior composition explorer.
NICMOS – (HST) Near Infrared Camera and Multi-Object Spectrometer.
NIVR – Netherlands Agency for Aerospace Programs
NM – nano-meter
Nova – transient astronomical event involving a bright new star that fades over time.
NRC – (Canada) National Research Council.
NRL – Naval Research Laboratory, Washington, D.C.
NSSDC -National Space Science Data Center.
NSF – (U.S.) National Science Foundation.
NSSC – NASA Standard Spacecraft Computer, an 18-bit flight computer.
NuSTAR - Nuclear Spectroscopic Telescope Array
OAO – Orbiting Astronomical Observatory.
Orbit – the path of one body around another, that are linked by gravity.
OSSE - Oriented Scintillation Spectrometer Experiment (CGRO).
OTE – optical telescope element (JWST)
PAMELA - Payload for Antimatter Matter Exploration and Light-nuclei Astrophysics.
Parsec – parallel second of arc, unit of length, about 3.26

light years.
PCA – proportional counter array.
PDR – preliminary design review.
Pera – nearest point to the primary, in an orbit.
pet – proton-electron telescope
Peta – 10^{16}
PHA – potentially hazardous asteroids
PHO – potentially hazardous objects
Photon – quantum particle of the electromagnetic field, zero mass, moves at speed of light.
Planet – a body orbiting a star.
Planetary disk – debris disks around a star.
Pleides – an open star clusterPHA
Pulsar – highly magnetized rotating neutron star of white dwarf.
Quark – an elementary particle.
Quasar – quasi-stellar object, galactic nucleus.
RAE - Radio astronomy explorer.
RC – High resolution camera (Chandra).
Red Shift – an apparent shift of electromagnetic radiation toward an increasing wavelength due to the doppler effect.
Ring system – a disk of solid material around a planet.
Rhessi - Ramaty High-Energy Solar Spectroscopic Imager.
Rockoon – a rocket taken to altitude by a balloon and then launched.
Rogue planet – planet not associated with a star.
ROSAT (short for Röntgensatellit), X-ray in German.
RTG – radioisotope thermoelectric generator.
RXTE – Rossi X-ray Timing Explorer.

Sampex - Solar Anomalous and Magnetospheric Particle Explorer.
SAO - Smithsonian Astrophysical Observatory.
SAS - Small Astronomy Satellite. (NASA)
Scintillation – variations in apparent brightness.
SD – scattered disk, a distant ring of smaller solar system objects, beyond Neptune
SDO – scattered disk object; Solar dynamics observatory
SERC – Science and Engineering Research Council (U.K.)
SES (Nasa-GSFC) space environment simulator.
SETI – search for extra-terrestial intelligence.
Seyfert galaxy – galaxy with quasar-like nuclei, about 10% of all known galaxies.
SGR – soft gamma ray repeater; emits large bursts of gamma-rays and X-rays at irregular intervals.
SIM – Science Instrument Module; space interferometry mission
SIRTF - Space Infrared Telescope Facility, renamed Spitzer.
SMEX – Small Explorer program (NASA)
SMM – Solar Maximum Mission.
SN – (NASA) space network.
SNSA - Swedish National Space Agency.
SOFIA – NASA Stratosphere Observatory for Infrared Astronomy.
Soft x-ray – energies below 5 kev
Solar flare – a sudden rapid emission of electrons, ions, and atoms from a star.
Solar System – A star and its associated planets and such.
Solar wind – stream of charged particles emitted from a

star's upper atmosphere.
Solo – (ESA) solar orbiter
Solstice – day of the shortest or longest period of daylight.
SRC - Science Research Council (UK)
SRON – Netherlands Institute for Space Research.
SSC – Swedish Space Corporation.
SSR – solid state recorder.
STOL – system test oriented (computer) language.
Strangelet – a hypothetical particle, made of up, down, and strange quarks.
STS – Space Transportation System (Shuttle).
StScI – Space Telescope Science Institute (JHU)
Supernova – a massive explosion of a star, at its end of life.
SWG – Science Working Group.
SWOOPS -Solar Wind Observations Over the Poles of the Sun.
TBD – to be determined.
TDRS – Tracking and Data Relay Satellite.
TDRSS – Tracking and Data Relay Satellite System.
TESS - Transiting Exoplanet Survey Satellite.
Tidal lock – where the same side of a object always faces the primary it is orbiting.
TLP -transient lunar phenomena.
TNO – Trans-Neptunian objects.
Trojan – minor planet that shares an orbit with one of the larger planets.
TRW - Thompson Ramo Wooldridge, Aerospace Company.
UK – United Kingdom, England.

USAF – United States Air Force.
USRA – Universities Space Research Association.
UV – ultraviolet, 19 nm to 400 nm wavelength.
WFIrST - Wide Field Infrared Survey Telescope.
VIM – Voyager Interstellar Mission.
Virtual telescope - several robotic telescopes, remotely available in real-time over the Internet.
WFI – Wide Field Instrument
Wfirst - Wide Field Infrared Survey Telescope
White dwarf – very dense remnant of a stellar core.
WISE - Wide Field Infrared Survey Explorer.
WMAP - Wilkinson Microwave Anisotropy Probe.
Wmops - WISE Moving Object Processing Software.
WUPPE - Wisconsin Ultraviolet Photo-Polarimeter Experiment.
XMM - X-ray Multi-Mirror Mission.
X-ray - 0.1 to 10 nanometer wavelength.
X-ray binary (star) – binary star, emitting x-rays.
XRCF – X-ray & cyrogenic facility (MSFC)
YSO – young stellar objects.
Zombie-sat – dead satellite, in orbit.

Bibliography

Armus, L., Reach, W. Y. "The Spitzer Space Telescope: New Views of the Cosmos: Proceedings of a Meeting Held in Pasadena, California, USA 9-12 November 2004," (Astronomical Society of the Pacific Conference Series), 2006

Associated Press, *The Hubble Space Telescope: A Universe of New Discovery*, 2015, ISBN-1633530469.

Baker, David NASA *Hubble Space Telescope - 1990 onwards (including all upgrades): An insight into the history, development, collaboration, construction and role of ... space telescope* (Owners' Workshop Manual), 2015, ISBN-100857337971.

Code, Arthur D. *OAO [Orbiting Astronomical Observatory]: Two Years of Scientific Achievement*, 1970, ASIN-B001J4X9AO.

Copernicus, Nicolaus *On the Revolutions of the Heavenly Spheres,*
1543, translated form Latin to English, ASIN-B01MS8TGOV.

Devorkin, David H.; Smith, Robert *Hubble: Imaging Space and Time*, 2008, National Geographic, ISBN-1426203225.

Devorkin, David H. *The Hubble Cosmos: 25 Years of*

New Vistas in Space, 2015, National Geographic, ISBN-9781426215575.

Dickinson, Terence *Hubble's Universe: Greatest Discoveries and Latest Images*, 2017, ISBN-9781770859975.

Edwards, Owen; Levay, Zoltan *Expanding Universe: Photographs from the Hubble Space Telescope*, 2015, ISBN-3836549220.

Ellerbroek, Lucas, *Planet Hunters: The search for extraterrestrial life,* 2017, ASIN-B073S986GV.

English, Neil S*pace Telescopes - Capturing the Rays of the Electromagnetic Spectrum,* Springer, 2017, ISBN-9783319278124.

NASA, *NASA's Great Observatories*, 2013, ASIN-B00DJUALH0.

NASA, "NASA Astrophysics Missions: Reviews of Operating Missions - Hubble Space Telescope, Chandra X-ray Observatory, Fermi Gamma-ray Telescope, Kepler, Planck, Suzaku, Swift, Warm Spitzer, XMM-Newton." 2012, ASIN-B007SH9FCM.

Perryman, Michael *The Exoplanet Handbook*, 1st Edition, Cambridge University Press, 2014, ISBN-1107668565.

Pesnell, W. D., B. J. Thompson, and P. C. Chamberlin,*The Solar Dynamics Observatory (SDO)*, Solar Physics, 2012. avail: http://adsabs.harvard.edu/abs/2012SoPh..275....3P

Reach, W. T. ; Armus, L. "The Spitzer Space Telescope: New Views of the Cosmos: Proceedings of a Meeting Held in Pasadena, California, USA 9-12 November 2004" (Astronomical Society of the Pacific Conference Series), 2006, ISBN-1583812253.

Rieke, George H. *The Last of the Great Observatories: Spitzer and the Era of Faster, Better, Cheaper at NASA*, 2006. University of Arizona Press, ISBN-0816525587.

Seager, Sara *Exoplanets*, 2011, University of Arizona Press, ISBN-0816529450.

Shayler, David J.,Harland, David M. The Hubble Space Telescope: From Concept to Success (Springer Praxis Books, 2015, ISBN-1493928260.

Summers, Michael *Exoplanets: Diamond Worlds, Super Earths, Pulsar Planets, and the New Search for Life beyond Our Solar System,* 2017, Smithsonian Books, ASIN-B01HA426MS.

Tasker, Elizabeth *The Planet Factory: Exoplanets and the Search for a Second Earth*, 2017, 1st Edition, ASIN-B01M7TMUTR.

Tucker, Wallace H. *Chandra's Cosmos: Dark Matter, Black Holes, and Other Wonders Revealed by NASA's Premier X-Ray Observatory,* 2017, ISBN-1588345874.

U. S. Government, *Complete Guide to the Kepler Space Telescope Mission and the Search for Habitable Planets and Earth-like Exoplanets - Planet Detection Strategies, Mission History and Accomplishments,* 2013, ASIN-B00CZE0E28.

Voit, Mark Hubble *Space Telescope: New Views of the Universe,* 2000, ISBN-0810929236.

Resources

- Central Operation of Resources for Educators, https://www.nasa.gov/audience/foreducators/CORE-Redirect.html

- NASA Educator Resource Centers, https://www.nasa.gov/offices/education/programs/national/ercn/home/index.html

- http://www.Spacescience.mission.gov/missions

- http://hubblesite.org/the_telescope/hubble_essentials/quick_facts.php

- http://hubblesite.org/ebooks

- Webb site: http://webbtelescope.org/

- SAS-B - https://nssdc.gsfc.nasa.gov/nmc/spacecraftDisplay.do?id=1972-091A

- https://www.space.com/5628-nasa-envisions-huge-lunar-telescope.html

- Lunar Observatory - http://www.canadensys.com/programs/ilo-1/

- PROSPECTS FOR NEAR ULTRAVIOLET

ASTRONOMICAL OBSERVATIONS FROM THE LUNAR SURFACE- LUCI, New Views of the Moon 2 – Asia 2018 (LPI Contrib. No. 2070). Avail: https://www.hou.usra.edu/meetings/newviews2018/pdf/6041.pdf

- https://www.nasa.gov/mission_pages/nustar/news/nustar20120727.html.

- "Complete Guide to NASA's James Webb Space Telescope (JWST) Project - Spacecraft, Instruments and Mirror, Science, Infrared Astronomy, GAO and Independent Review Reports, Congressional Hearings, 2011" 20178, ISBN-1549878212.

- www.planetary.org

- https://heasarc.gsfc.nasa.gov/docs/

- https://www.nasa.gov/content/goddard/parker-solar-probe

- Gunter's Space Page - https://space.skyrocket.de/

- http://www.brite-constellation.at

- (WMAP data) https://lambda.gsfc.nasa.gov/product/map/dr5/m_products.cfm

- http://parkersolarprobe.jhuapl.edu/index.php#spacecraft

- http://www.spacefab.us/space-telescopes.html

- National Research Council; Division on Engineering and Physical Sciences; Board on Physics and Astronomy;Space Studies Board; Committee for a Decadal Survey of Astronomy and Astrophysics
"New Worlds, New Horizons in Astronomy and Astrophysics" (2010),
https://www.nap.edu/catalog/12951/

- wikipedia, various.

If you enjoyed this book, you might find something else from the author interesting as well.

Stakem, Patrick H. *16-bit Microprocessors, History and Architecture*, 2013 PRRB Publishing, ISBN-1520210922.

Stakem, Patrick H. *4- and 8-bit Microprocessors, Architecture and History*, 2013, PRRB Publishing, ISBN-152021572X,

Stakem, Patrick H. *Apollo's Computers,* 2014, PRRB Publishing, ISBN-1520215800.

Stakem, Patrick H. *The Architecture and Applications of the ARM Microprocessors,* 2013, PRRB Publishing, ISBN-1520215843.

Stakem, Patrick H. *Earth Rovers: for Exploration and Environmental Monitoring,* 2014, PRRB Publishing, ISBN-152021586X.

Stakem, Patrick H. *Embedded Computer Systems, Volume 1, Introduction and Architecture*, 2013, PRRB Publishing, ISBN-1520215959.

Stakem, Patrick H. *The History of Spacecraft Computers from the V-2 to the Space Station*, 2013, PRRB Publishing, ISBN-1520216181.

Stakem, Patrick H. *Floating Point Computation*, 2013, PRRB Publishing, ISBN-152021619X.

Stakem, Patrick H. *Architecture of Massively Parallel Microprocessor Systems*, 2011, PRRB Publishing, ISBN-1520250061.

Stakem, Patrick H. *Multicore Computer Architecture*, 2014, PRRB Publishing, ISBN-1520241372.

Stakem, Patrick H. *Personal Robots*, 2014, PRRB Publishing, ISBN-1520216254.

Stakem, Patrick H. *RISC Microprocessors, History and Overview,* 2013, PRRB Publishing, ISBN-1520216289.

Stakem, Patrick H. *Robots and Telerobots in Space Applications*, 2011, PRRB Publishing, ISBN-1520210361.

Stakem, Patrick H. *The Saturn Rocket and the Pegasus Missions, 1965,* 2013, PRRB Publishing, ISBN-1520209916.

Stakem, Patrick H. *Visiting the NASA Centers, and Locations of Historic Rockets & Spacecraft,* 2017, PRRB Publishing, ISBN-1549651205.

Stakem, Patrick H. *Microprocessors in Space*, 2011, PRRB Publishing, ISBN-1520216343.

Stakem, Patrick H. Computer *Virtualization and the Cloud*, 2013, PRRB Publishing, ISBN-152021636X.

Stakem, Patrick H. *What's the Worst That Could Happen? Bad Assumptions, Ignorance, Failures and Screw-ups in Engineering Projects*, 2014, PRRB Publishing, ISBN-1520207166.

Stakem, Patrick H. *Computer Architecture & Programming of the Intel x86 Family*, 2013, PRRB Publishing, ISBN-1520263724.

Stakem, Patrick H. *The Hardware and Software Architecture of the Transputer*, 2011, PRRB Publishing, ISBN-152020681X.

Stakem, Patrick H. *Mainframes, Computing on Big Iron*, 2015, PRRB Publishing, ISBN- 1520216459.

Stakem, Patrick H. *Spacecraft Control Centers*, 2015, PRRB Publishing, ISBN-1520200617.

Stakem, Patrick H. *Embedded in Space,* 2015, PRRB Publishing, ISBN-1520215916.

Stakem, Patrick H. *A Practitioner's Guide to RISC Microprocessor Architecture*, Wiley-Interscience, 1996, ISBN-0471130184.

Stakem, Patrick H. *Cubesat Engineering*, PRRB Publishing, 2017, ISBN-1520754019.

Stakem, Patrick H. *Cubesat Operations*, PRRB Publishing, 2017, ISBN-152076717X.

Stakem, Patrick H. *Interplanetary Cubesats*, PRRB Publishing, 2017, ISBN-1520766173 .

Stakem, Patrick H. Cubesat Constellations, Clusters, and Swarms, PRRB Publishing, 2017, ISBN-1520767544.

Stakem, Patrick H. *Graphics Processing Units, an overview*, 2017, PRRB Publishing, ISBN-1520879695.

Stakem, Patrick H. *Intel Embedded and the Arduino-101, 2017,* PRRB Publishing, ISBN-1520879296.

Stakem, Patrick H. *Orbital Debris, the problem and the mitigation*, 2018, PRRB Publishing, ISBN-*1980466483*.

Stakem, Patrick H. *Manufacturing in Space*, 2018, PRRB Publishing, ISBN-1977076041.

Stakem, Patrick H. *NASA's Ships and Planes*, 2018, PRRB Publishing, ISBN-1977076823.

Stakem, Patrick H. *Space Tourism*, 2018, PRRB Publishing, ISBN-1977073506.

Stakem, Patrick H. *STEM – Data Storage and Communications*, 2018, PRRB Publishing, ISBN-1977073115.

Stakem, Patrick H. *In-Space Robotic Repair and Servicing*, 2018, PRRB Publishing, ISBN-1980478236.

Stakem, Patrick H. *Introducing Weather in the pre-K to 12 Curricula, A Resource Guide for Educators*, 2017, PRRB Publishing, ISBN-1980638241.

Stakem, Patrick H. *Introducing Astronomy in the pre-K to 12 Curricula, A Resource Guide for Educators*, 2017, PRRB Publishing, ISBN-198104065X.
Also available in a Brazilian Portuguese edition, ISBN-1983106127.

Stakem, Patrick H. *Deep Space Gateways, the Moon and Beyond*, 2017, PRRB Publishing, ISBN-1973465701.

Stakem, Patrick H. *Exploration of the Gas Giants, Space Missions to Jupiter, Saturn, Uranus, and Neptune*, PRRB Publishing, 2018, ISBN-9781717814500.

Stakem, Patrick H. *Crewed Spacecraft*, 2017, PRRB Publishing, ISBN-1549992406.

Stakem, Patrick H. *Rocketplanes to Space*, 2017, PRRB Publishing, ISBN-1549992589.

Stakem, Patrick H. *Crewed Space Stations*, 2017, PRRB Publishing, ISBN-1549992228.

Stakem, Patrick H. *Enviro-bots for STEM: Using

Robotics in the pre-K to 12 Curricula, A Resource Guide for Educators, 2017, PRRB Publishing, ISBN-1549656619.

Stakem, Patrick H. *STEM-Sat, Using Cubesats in the pre-K to 12 Curricula, A Resource Guide for Educators*, 2017, ISBN-1549656376.

Stakem, Patrick H. *Embedded GPU's*, 2018, PRRB Publishing, ISBN- 1980476497.

Stakem, Patrick H. *Mobile Cloud Robotics*, 2018, PRRB Publishing, ISBN- 1980488088.

Stakem, Patrick H. *Extreme Environment Embedded Systems,* 2017, PRRB Publishing, ISBN-1520215967.

Stakem, Patrick H. *What's the Worst, Volume-2*, 2018, ISBN-1981005579.

Stakem, Patrick H., *Spaceports*, 2018, ISBN-1981022287.

Stakem, Patrick H., *Space Launch Vehicles*, 2018, ISBN-1983071773.

Stakem, Patrick H. *Mars*, 2018, ISBN-1983116902.

Stakem, Patrick H. *X-86, 40th Anniversary ed*, 2018, ISBN-1983189405.

Stakem, Patrick H. *Lunar Orbital Platform-Gateway*, 2018, PRRB Publishing, ISBN-1980498628.

Stakem, Patrick H. *Space Weather*, 2018, ISBN-1723904023.

Stakem, Patrick H. *STEM-Engineering Process*, 2017, ISBN-1983196517.

Stakem, Patrick H. *Space Telescopes,* 2018, PRRB Publishing, ISBN-1728728568.

Stakem, Patrick H. *Exoplanets*, 2018, PRRB Publishing, ISBN-9781731385055.

Stakem, Patrick H. *Planetary Defense*, 2018, PRRB Publishing, ISBN-9781731001207.

Patrick H. Stakem *Exploration of the Asteroid Belt*, 2018, PRRB Publishing, ISBN-1731049846.

Patrick H. Stakem *Terraforming*, 2018, PRRB Publishing, ISBN-1790308100.

Patrick H. Stakem, *Martian Railroad,* 2019, PRRB Publishing, ISBN-1794488243.

Patrick H. Stakem, *Exoplanets,* 2019, PRRB Publishing, ISBN-1731385056.

Patrick H. Stakem, *Exploiting the Moon,* 2019, PRRB

Publishing, ISBN-1091057850.

Patrick H. Stakem, *RISC-V, an Open Source Solution for Space Flight Computers,* 2022, PRRB Publishing, ISBN-1796434388.

Patrick H. Stakem, *Arm in Space*, 2019, PRRB Publishing, ISBN-9781099789137.

Patrick H. Stakem, Search for *Extraterrestrial Life*, 2019, PRRB Publishing, ISBN-978-1072072188.

Stakem, Patrick H. Submarine Launched Ballistic Missiles, 2019, ISBN-978-1088954904.

Patrick H. Stakem, *Space Command*, Military in Space, 2019, PRRB Publishing, ISBN-978-1693005398.

History & Future of Cubesats, ISBN-978-1986536356.

Robotic Exploration of the Icy Moons of the Ice Giants, by Swarms of Cubesats, ISBN-979-8621431006.

Swarm Robotics, 2021, ISBN-979-8534505948.

Introduction to Electric Power Systems, 2021, ISBN-979-8519208727.

Powerships, Powerbarges, Floating Wind Farms: electricity when and where you need it, 2021, PRRB Publishing, ISBN-979-8716199477.

Centros de Control: Operaciones en Satélites del Estándar CubeSat (Spanish Edition), 2021, ISBN-979-8510113068.

The Artemis Missions, Return to the Moon, and on to Mars, 2021, ISBN-979-8490532361.

James Webb Space Telescope A New Era in Astronomy, 2021, ISBN-979-8773857969.

Exploration of Icy Moons and Ocean Worlds, coming in 2022.

Back to the Moon, Boots on the Regolith, coming in 2022.

Robots Underground, coming in 2022.

www.ingramcontent.com/pod-product-compliance
Lightning Source LLC
Chambersburg PA
CBHW031438210526
45464CB00005B/2249